MW00964039

~~~~~~~~~~~

Adventures In Saying YES To God!

~~~~~~~~~~~

Kel Steiner

with Andrew Kooman

With a foreword from Loren Cunningham
Founder of Youth With A Mission

To order copies of this book, or to inquire regarding Kel's availability to speak in your conference, training center or church, contact: kelsteinerbook@gmail.com

YES... GOD!
Registered with the Library of Congress Serial Number: 1-791122621

About: A young farmer's son from New Zealand finds himself in exciting and sometimes frightening situations in the far off jungles and cities of Asia. His diverse travels take him to the slums and red light districts of Asia's mega cities, remote islands of the Pacific and the world's most modern cities. He describes the thrill and excitement of hearing God's voice, the miracle of leading lost tribes to Christ and discipling the nations. Kel's story reveals what a modern missionary's life is like, both the fun and the challenges of discipleship in the 21 century.

Author: Steiner, Kel [1947 -]

For a free catalog of books and materials by YWAM authors, or to receive a GO Manual of YWAM ministries around the world that you can become involved in, write or call:
YWAM Publishing, P.O. Box 55787, Seattle, WA 98155
1 (425) 771-1153 or 1 (800) 922-2143
http://www.ywampublishing.com

Unless other wise notated, Scripture quotations are taken from the New King James Version of the Bible. (NKJV)

Dedication

This book is written with the hope it will inspire and challenge the emerging generation to live for a cause greater than themselves.

Acknowledgements

Where do you begin when you owe so much to so many precious people whose lives have decisively influenced you and in many meaningful ways contributed to the substance and completion of this book?

My parents, June and Bert Steiner: for a wonderful childhood.

Alex and Wynn Laskey: for planting the seed of missions in my young heart at the Welcome Bay Sunday School, in New Zealand.

Ross and Margaret Tooley: who taught me so much about evangelism.

Loren and Darlene Cunningham: for founding Youth With A Mission, and modeling and mentoring the Christ life, thus providing an exciting mission in which to serve God.

Andy Elliott: transcribing the early script.

Susan Allan, Melanie Hurlbut, Sean Sanborn and my son John-Michael and daughter Karissa Steiner: for editing.

Helen Dean of http://www.helendeandesign.com/: for cover design, sketches and book layout.

Andrew Kooman, for his masterful writing skills.

My wonderful wife and three children, who journeyed with me to the far corners of the earth.

Foreword by Loren Cunningham

God created you for adventure! And He filled this world with such astonishing diversity that it is guaranteed your adventure will be different from every other person's on Earth. Adventure is much more than unexpected encounters and unplanned routes, it is a place of learning, risk-taking, and determination. It is never giving up. If you accept God's plan for your life, then you will walk into a world that you were unaware existed. And though you might not yet know it, this is the world you want. Fulfilling God's plan for your life is the world you were born to live in.

Whether your adventure is in accounting, or along the muddy slopes of a jungle mountainside where tribal groups are facing you and you don't know how to react... well, Kel Steiner has wisdom for you!

Kel is the kind of guy that wants to be out there in the jungle. He wants to be in a place where he can be challenged. And as you read these stories—no matter to what adventure God calls you to—you will learn principles that are universal. Adopting the principles found in this book will allow God to uniquely prove His faithfulness and commitment to you.

But first you must choose God's will—you have to choose God's adventure. When you do, you will notice, as in 2 Chronicles 16:9, that the Lord is looking for those whose hearts are lined up with His own so that He can be strong on your behalf. This is a key underlining principle of Kel's book. As I read it, I could only say "wow": it was like walking through memory lane, seeing things from the past that work in the present, and that will continue to work in the future.

Don't sell yourself short. Don't make your life dull and boring. Be ruined for the ordinary by walking with God in the adventure He has planned for you.

Loren Cunningham - Founder of Youth With A Mission

Foreword By Tom Hallas

This book should have been written way before this time. The delay, however, has proven how very real the stories are. Kel Steiner documents a life of obedience to the still, small voice of God and what can be the long term outcomes of that lifestyle. His stories are remarkable and yet ongoing. You will be encouraged, challenged and blessed.

Tom Hallas
Asia/Pacific Field Leader, Youth With A Mission

Foreword by John Dawson

What an adventure. What a life. What a refreshing and uplifting book. I could not put it down. Kel's amazing story will help you to dream again. There are godly ambitions almost forgotten in the hearts of many believers. If you want a guide to a higher, more joyous future, this book is a good place to begin.

John Dawson
President Emeritus, Youth With A Mission

Table of Contents

Introduction: I Said... Yes

I could have said no, it's true. We all can. We all do. We say no because of fear or uncertainty; because when you say yes to God, there are other things we have to let go of—our plans, and sometimes the dreams we have of what we want our lives to be.

I could have said no, but I didn't.

Though I don't think about it often, I am aware that the whole thrilling adventure of my life could have unfolded in a completely different way. I am also aware that God could have done things differently. He could have captured my young heart in a hundred different ways, but stories were all He needed. These were stories of brave men who gave up everything they had—stories of men who traveled the world, who left everything they knew, took great risks for God, and who ended up experiencing incredible adventures.

God could have inspired me when I was young to be a missionary in other ways, but He did it through stories of great missionaries. Hudson Taylor. CT Studd. William Carey. These men were heroes to me when I was a boy; they stood like towering figures in my imagination. These were men whose lives I aspired to emulate—men who changed the world by simply responding in obedience to the call of God.

God's invitation for me to become a missionary transformed my life's story into one that far surpassed any hope or dream for adventure that I could have mustered in my youth.

That's the story I want to tell you. Consider it my way of saying that God still works in amazing and mysterious ways. Because of His great love for the world, God calls ordinary people like you and me to work for him.

I should be honest with you (maybe even warn you, or possibly your parents!): I'm writing this book because I'm convinced that God has an exciting plan for you. I am convinced that God is calling you into an adventure that will surpass any plan you might have for your own life. How do I know this is true? Because it's been my own experience and the experience of countless others who have risked saying Yes God!

This book is an account of my life in missions—of what happened when I said Yes to God.

Chapter 1: Wuh Tribe

The words were still fresh in my mind though it had been days since I heard them. It seemed impossible that I had heard them at all; but the deeper we went into the jungle, the more loudly they replayed in my mind. “Jones to Command, you won’t believe what I’ve just seen.” The pilot’s voice was clear in my memory. It had come through the radio days earlier by some miracle and full of static—scratchy—but distinctly audible. Was what I heard real?

I would soon find out. Ivan, who followed directly behind me, had no doubts. Neither did the two young Filipino girls who walked closely behind us. Carmen and Vangie walked with a resolve like I had never seen, just as they promised they would, showing no signs of fatigue. Together we ventured deeper into the trees. With each step we left civilization and the familiar world further behind and trekked deeper into the humid rain forest. They were following me.

New Zealand now seemed like a distant memory. What an amazing detour this life was from the life I had imagined for myself as a boy! Abruptly my nostalgia was interrupted by the sound of an American military jet flying overhead.

“An F-5E Tiger!” Ivan shouted as our small expedition looked up through the trees. “I’d know that sound anywhere!”

It was gone before we could see it. The booming noise of the jet was like one massive piece of paper being ripped across the length of the sky.

“It must be another military jet doing reconnaissance in the area,” I said, and set my sights back on the jungle that was ahead of us.

One of the girls asked, “How much further do you think we’ll walk today? Soon it will be dark.”

“Point taken” I replied. “It could be dangerous to sleep out here. Who knows what animals might be roaming the jungle floor tonight in search of something to eat!"

"Those are the least of my concerns," said Ivan, smiling. "I'm more worried about an indigenous tribe bumping into us and not being pleased to find trespassers sleeping on their land."

"We'll keep walking," I said, "at least for another hour." I pulled out the map from my pocket and unfolded its tattered pages. "The officer who dropped us at the edge of the jungle said we would run into a village not far from here."

We continued to walk, weary and fatigued. My back ached from the weight of my pack, which was filled to bursting with food and medical supplies. My legs burned from the steep trek. How many hours had we been in the hot jungle heat? But the words of the pilot urged me on, motivating me to find the mysterious tribe we were led to believe lived high up on a mountain ridge. We were seeking a tribe that perhaps no one but that pilot had ever laid eyes on.

How long we had walked since the plane surprised us overhead I could not say. We came to a point in the trail that edged the side of a mountain. Ahead of us was a steep incline, and across from the rock face, vibrant jungle surrounded us. I stopped dead in my tracks. Ivan bumped into me.
"Kel, are we stopping now?" I held my finger to my mouth and he stopped talking.

We were at the base of a sharp incline, and at the top of the hill stood a man. He was naked except for a covering of vines that were wrapped around his loins. Across his chest was a strap that held a collection of arrows on his back a full foot taller than his head; and in his right hand, he held a bow that was two feet taller than himself.
The four of us stood closely together in a group. We looked up at the warrior and he looked down at us. A long moment passed, quiet except for the buzzing of insects in the jungle and the thumping of our hearts in our throats, our ears, between our temples, and in our chests. Did he mean us harm?

"You need to say something," whispered one of the girls, breaking the silence.

"You know the mountain languages, not me," I urgently whispered back to them, keeping my eyes fixed on the man.

"It's not us he's interested in, Kel, it's you."

I looked up the hill and saw that she was right. The warrior's gaze was fixed directly on me. Because I stood at the head of our small group, he gathered that I was leading them. No one else was of interest to him. And the rest of our journey toward the lost tribe—perhaps even our very lives—depended on what I would say to this man, standing above us, bow in hand.

For the life of me, in that moment, I could not think of a thing to say.

Chapter 2: Light Shines in the Darkness

It was pitch black, that's what I remember most. The smell of hay surrounded us. The center of the barn was warm with the bodies of my siblings —all of us cramped into the tight space with no wiggle room. The small pocket of air in the barn's center was filled with the sound of our breathing. And then the "Ffwt ffwt" of a match being struck was heard as it suddenly burst into a small flame.

"Quick, Kel, the candle!" I could see my brother Les mouth the words in the eerie darkness as the match quickly burned. "Hurry!"

I pulled the candle out of my pocket and held the wick near to the burning match. The small flame was inching toward Les' fingers, and I could see the urgency in his usually stoic eyes. The wick took flame and Les quickly shook out the match.

"Careful!" Ray urged. "The hay!"

All around us were bales of hay packed one on top of the other, from the outer walls of the barn toward the center. If the bales collapsed they would fall inward, smothering us as we huddled in the center of the barn. We would be buried alive. One careless move with the match, or the candle, and the whole barn would go up in smoke and we would be trapped in the middle of an inferno.

"So, why are we meeting today?" I asked Les, my older brother.

The four of us, Les (age fourteen), Ray (age twelve), Herb (age eight), and I (age six), sat in a tight circle, holding our knees to our chests as Les took the candle from me and brought it close to his face. He spoke in a near whisper. We all leaned in toward the light.

"Today I will tell you the story of our family, and how we came to New Zealand."

I smiled at Ray and we all leaned in closer to Les.

"Every story starts somewhere and ours began like so many others. Two people fell in love and got married."

"Not a love story!" Herb moaned.

Les ignored the complaint and continued. "You may know that Dad's family came across the ocean to New Zealand from Switzerland when he was a young boy, only seven years old. But did you know that all eight of us are lucky to be alive today?"

"Did Dad almost die?" Herb asked, suddenly interested in Les' tale.

"No," Ray said. "Come on Les, don't make up stories."

"I'm not," Les assured.

"If you're going to tell the same story of how Dad had to work the land from a young age, and get up around 5:30 every morning to go and get the cows for milking, then you may as well stop. We already know that story," Ray said.

"I'm not."

"Dad's already told us a million times that every day he'd milk one hundred and twenty cows and then hurry home to change his clothes, gulp down a breakfast, and get to school on time." Ray leaned back and rested against a hay bale.

The Great Depression reached far and wide. It dried up economies around the world, including New Zealand's. Families could rarely afford to hire workers, so every one in the family had to chip in to eek out an existence. We had all heard the story before. Dad had to leave school at fourteen to work on the family farm just to help make ends meet. It was information I already knew. I also knew that my Dad worked the farm without a tractor, since they were not yet invented. Instead, he had to use horses to pull the heavy cans full of cream from the milking shed to a stand on the side of the dusty gravel road. From there they would be collected and transported to a factory to be processed into butter, cheese and milk powder.

"You know that Dad worked hard," Les said, "but did you know that after all those years working the land with Grandpa, that he never got a penny?" It was Les' turn to lean into the hay. He kept the candle close to his chest so that Ray, Herb and I were shrouded in darkness. Squinting in the dark, Les smiled and continued.

"And, did you know that the Catholic priest put a curse on Mum and Dad when Dad refused to break off his engagement?"

Small gasps of surprise from the three of us filled our small cave, and the candle flickered from the sudden intake of air.

"He worked long hours, from sunrise to sunset. One day in the nearby countryside a woman caught his eye. When Dad told Grandpa that he wanted to marry Mum, Grandpa disapproved."

"Why?" I asked.

"Because Mum was a Protestant," Les said.

"What's a Protestant?" Herb asked. "It sounds no good."

Les laughed. "It's not bad. It's a kind of Christian who goes to a certain kind of church."

"Well what's wrong with that?" Herb asked.

"Nothing," Les assured. "It's a good thing, really. It's just that Grandpa and his family were Catholics from a different church, Herb, and in those days, Catholics and Protestants didn't get along. Grandpa said 'If you marry her, we'll have nothing more to do with you!' The family even called the priest to convince Dad not to marry Mum, but he had already made his decision. He was in love with Mum, and couldn't imagine marrying anyone else. On the third and final visit to Dad, the priest said, 'If you marry this girl, your wife will be barren.'"

"What does barren mean?" I asked.

"It means you can't have kids," Ray said.

"But she did!" Les chuckled.

"Eight of us!"

We all laughed. Les leaned forward again and brought the candle close to our faces. One by one he looked at us seriously. "Dad got married when he was twenty-six. He had worked the farm since he was fourteen. Grandpa refused to pay him the wages he had banked all those years on the farm—more than ten years of work—and so Dad didn't receive a single cent for all his hard labor!"

We sat in silence for a moment thinking about this.

"Is that why we're poor?" Ray asked.

"That's why we're poor." Les said.

"How come Dad let those missionaries on our land, then?" I asked, surprised in light of Les' story, that Dad allowed a religious couple to park their caravan on our farmland. The Kiwi couple were traveling Pentecostal missionaries, who often talked about the fire of God. Alan Fawcett was a preacher and he and his wife Gladys had started to teach us the Bible in our home every week.

"What's strange about that?" Les asked.

"Well, don't you think dad would want nothing to do with religion if Grandpa treated him so badly?" Ray asked.

Les shrugged his shoulders. "Why, do you wish they weren't here?"

"No," I mused. "I'm glad he invited them. It's nice to have visitors, and I'm glad the preacher's wife plays the piano accordion."

"Like the preacher says, light shines in the darkness." With that, Les blew out the candle, and chuckled as the rest of us yelled out in the sudden black out. We couldn't see a thing. All of us groped in the dark as we made our way slowly through the tiny tunnel in the hay bales, back to the bright and pleasant world of the outdoors to do our chores for the day on our farm.

Chapter 3: Afire with Faith

Just one spark from Les' match could have set our barn ablaze. Perhaps it was a miracle that kept the flames from consuming us as young kids when we met there secretly in the dark. Instead, by God's grace, it was my thirteen-year old heart that was set on fire. The visiting missionary couple who lived on our farm introduced me to the love of God. Their teaching, their reading of the Bible, and the catchy worship songs they sang sparked faith in my heart. And the exciting stories they shared lit that faith ablaze.

When my dad bought a dairy farm in Welcome Bay—a region known for its kiwifruit, citrus, and apple orchards—we left our farm in Hinuera and moved to Tauranga. As we settled into our new home, my parents enrolled us at an interdenominational Sunday school that was run by Alec and Wynn Laskey (these two became my spiritual mum and dad). Mostly farm kids, like us, were in attendance there.

My first visit to the Sunday school was a happy surprise. "Pleased to meet you all," spoke a gentle voice. I had never heard that voice before, but it is a voice that I will never forget until the day I die. I turned around to see a woman smiling down at us. "My name is Mrs. Laskey and this is my husband, Mr. Laskey," she said pointing to a tall man in a brown suit, sporting a mustache. "Come on, take a seat," she said pointing to a semi circle of kids facing a stage.

As I looked around the room, I recognized many of the kids from school. Each weekday morning we would pile into a small green bus, which I always thought was better suited to the description of a swollen green insect, and travel to school. The bus driver was in sharp contrast to the neatly dressed and smiling Laskeys. He was a short little dumpling of a man, who hunched over the steering wheel chewing tobacco. The Laskeys were tall and regal, warm and kind. We all sat spotless in our Sunday best.

After introductions were over, and classes finished, we would all sing together. We then spent the remainder of the morning with our own age groups learning Bible stories. Before being dismissed, to our delightful surprise, Mr.

Laskey brought out a huge one-gallon glass jar of boiled lollies (our word for candies). We were all allowed to put our hand in and choose one. What a way to finish Sunday school! We were hooked from day one.

Stories that Changed My World

I don't recall what Mr. Laskey shared that first day at Sunday school, but I do remember watching the look of happiness on all my siblings' faces. Never had we felt so quickly welcomed or loved by strangers. The Laskey's faces shone with the warmth of the sun, and throughout all the years we attended the Sunday school, they taught us about Jesus. The beautiful way they lived their lives and interacted with each other, and all of us, was a great demonstration of godliness. The relationship they had with Jesus attracted us.

"Jesus loves you, all of you," the Laskeys would often say. It was words like these, and the kindness they showed Sunday after Sunday that captured my young heart with the love of God. There were over a hundred children who attended, and each week we would be assigned different memory verses to learn from scripture. I learned the verses every week, and by the end of a year, there were dozens of stickers beside my name as proof.

"Kel, you've done such a good job remembering Bible verses this year." Mrs. Laskey stood in front of the whole group of one hundred kids during the Sunday school anniversary. The hall was packed with parents for this occasion, we had all looked forward to it for weeks. It was the first time I was going to be awarded a prize. I could hardly wait for my name to be called out.

When my name was finally called, I walked up to the front of the room, beaming. Mr. Laskey handed me a book. "Here, this is for you. It's the story of a great man of faith who changed the world because of his love and obedience to God." He winked at me as my eyes lit up. "Be careful. It's an adventure story."

With that book my imagination and love for adventure was fueled by stories of missionary pioneers. Each year, as the Sunday school awards rolled by, many more books were handed out as prizes. God used all of these to captivate my heart.

I sailed with Hudson Taylor across the sea, watched him risk his life to work in China. I read in astonishment as William Carey taught himself Hindi, and then spent years in India translating the whole Bible into this complex foreign language. I read in disbelief as a terrible fire consumed all those years of work before it could ever be printed. Having lost all, he then started over, he translated it all again. And then there was CT Studd, blazing a trail through Africa, leaving his vast wealth and fame behind in England to preach the gospel.

The stories of these men seemed so exciting, so fulfilling: their death defying work, the sacrifices they made, the adventures they went on, and the souls that were saved. It seemed as though serving God launched people into the most exciting and wild adventures. I longed to do something for God that would help release people from spiritual darkness.

It was with an inspired imagination that I would return to the farm during the week. I imagined having adventures in other parts of the world as I jumped between hay bales and ran in the fields.

A Day of Decision

I continued to attend Sunday school, memorize scripture, and read mission adventures. However, it wasn't until I was thirteen that I made a public confession of my faith. Muri Thompson, a Maori Evangelist, was holding meetings at the Memorial Hall in Tauranga. At the time I didn't know what a public confession was, but after hearing him preach, there I was on my knees at the front of the hall. My heart was burning in my chest. I knew I was a sinner after he preached his message, I wanted God to deal with my life.

As I closed my eyes at the altar the image of a scroll flashed across my mind. What is that? I thought to myself, stunned at what I saw. I opened my eyes and looked around at the other people who had come to the front of the hall. All of them were on their knees experiencing God in their own way.

After surveying the room to see if anyone else was seeing what I was seeing, I closed my eyes again. Once more the scroll burned in my mind's eye. On the top of the scroll was written in bold letters: Lying to parents. My heart dropped. Why were the letters so big? I felt terrible, and knew, without having to ask anyone, that these words were exposing a sin in my own life. I opened my eyes and looked around the room. No one was looking at me. This was a moment between God and I, and I knew I had to confess my sin.

"I'm sorry, God, for lying to my parents" I muttered, my eyes still closed. Instantly, the scroll rolled up and disappeared. What a relief! But there was another. The scroll unrolled and at the top of it was the word "Stealing" as big and as bold as the words on the first scroll. My heart dropped again. I felt weak and ashamed of my sin.

"God, I'm sorry for this sin too," I said, truly convicted. When I confessed to God, this scroll rolled up too, and then another appeared. The sin in my life continued to appear in bold letters on separate scrolls. I kept repenting of the different sins that appeared on these scrolls until there were no more.

I don't know how long I stayed up front, watching these scrolls reveal themselves to me and then roll up, but when the last one disappeared I was finally right with God. What freedom I felt! Like a bird perched suddenly at the door of a cage that had swung wide open, my soul could no longer be contained, and was released. It could fly, higher and higher. I felt I had been set free like that caged bird, and my childhood sin was dealt with. With that freedom came a verse I remember to this day: "Be not entangled again with the sins of your youth." That day marked the public confession of my faith.

Through this defining experience, and the continued teaching of the Laskeys at Sunday school, my life's call into missions was planted like a seed in my heart. It was watered and grew as I memorized scripture, especially as I read the stories of missionaries. My heart became pregnant with this seed by the time I entered my adult years. Soon I would be pulled into a thrilling, lifetime adventure and serving God all over the world.

Chapter 4: The Olympics or Missions

The alarm went off again like it did every morning, but no matter how much I heard it, the trill, high pitched sound sent my heart racing. 4:00 AM, the red number read. "How come the mornings come so fast and early?" I moaned as I rolled my legs to hang over the edge of the bed. Though there was nothing more I wanted than a few more hours of sleep, I forced myself out of bed.

My bare feet touched the cool floor and I quickly dressed. As I hopped around in the cold, I could feel the same subtle pain in my shinbone that I had started to notice a number of days ago. I had been training for months, and working like a dog in between to earn money to save up for a farm of my own.

"You can do it Kel," I encouraged myself as I walked into the kitchen and grabbed a bowl of cereal. "Olympic champions must train hard."

I had another long day ahead of me. I went over the schedule in my head. First, I would have to get to the auction mart very early in the morning to help set-up for the day's auction. It was a well paying job that I landed soon after high school, and because of it I had saved up a handsome amount of money toward the purchase of the kiwi farm I had my eye on. I'd work until 5:00 PM. A solid twelve-hour day. I'd then head out to the running track.

Even more than I dreamt of having my own farm, I dreamt of being an Olympian. I had decided that I didn't want the world to pass me by like it seemed to pass by so many others. I had been drawn into athletics in high school, enjoying the challenge of pushing my body to its limits, and the exhilaration of a race. Following high school, I started to train on my own for the half-mile.

As I finished up my breakfast I walked out the door. A flare of pain shot up my leg again. "There's a price to pay for training," I said under my breath. "If you want to be a champion, you need to keep pushing yourself."

At that time, the best runners in the world were running the half-mile in around one minute and fifty seconds. I was consistently running under two minutes, but knocking a few more seconds off to get to a world-class time meant you not only had to have exceptional natural talent, but you had to be determined to endure hours of long, hard training sessions. I wasn't afraid of the training, the question in my mind was the exceptional talent bit!

I was twenty-one and anything seemed possible, even the dream to represent my country at the Olympic Games. I was willing to do whatever it took, and make the sacrifices necessary to achieve my dream.

"You look tired Kel," Erik, my boss said, as I showed up at work, just in time. "You still training?"

"Sure am," I said, proudly.

"How many hours did you train yesterday?"

"Same as everyday. Three!"

"Three hours of running? If I did that, I'd fall over dead. You're going to wear yourself out," he said.

Just like every other day, I worked hard, went to the track, and put in my hours of training before heading home. As usual I was tired beyond belief. After supper, I was completely exhausted. As I slipped under the covers and leaned to turn off the light beside my bed, I saw the same thing that made me feel guilty night after night: my Bible on the bedside table, collecting dust.

"God," I said pathetically, a ritual as consistent as my physical training after work, "you know that I love you. I'm sorry for not reading your Word again tonight. I'm just so tired."

I put my hands on the light switch, but couldn't shut out God any longer. My life was too busy, and I knew it. "Lord," I prayed, "You know how much I want to be a successful athlete. If you don't want my life to run along that track, then please put a stop to it."

I didn't feel totally relieved of my guilt, but I was able to turn out the light and go to sleep. I smiled in the dark. Tomorrow I could sleep in. I didn't have to get up for work until 5:30 AM!

I woke in the morning with renewed energy and a bounce in my step. I went through the motions of the day at work and then eagerly returned to the track to train. And that very day, God was faithful to answer my prayer. I broke my leg.

As I ran loops around the track, the same pain in my shin started to intensify. I decided to push through it. Within a few steps, however, the pain was too much. I hobbled to a stop and limped to my car. I knew I needed to see a doctor.

"Well, there's good news and there's bad news, son. Which do you want first?" the doctor asked after the x-ray.

"May as well hit me with the bad news, doctor," I said.

"Well, the bad news is that your leg is fractured."

"And the good news?" I asked, weakly.

"It's not all that severe. I'll bandage you up and you can be back training in no time."

True to his word, within three weeks my leg was healed, and I was ready to get back into it. With the doctor's clearance, I was soon out on the track, determined to pursue my dream of the Olympics. The training went well, and for a couple of months I enjoyed my busy routine of work and running. Despite my good intentions to spend time with God, I didn't find the time. Instead, every night when I fell into bed, my dusty Bible stared at me before I turned out the lights.

My recovery was so quick that I was able to convince myself that the break was a freak accident, a minor setback in my athletic dream. The training, however, wasn't to last very long. To the doctor's surprise, not to mention my own, I was soon in his office again. It was a cool afternoon one day after work, the best kind for a run, when the bone broke a second time.

"You're just one unlucky guy Kel," the doctor said, as he held the x-ray up to the light. "It shows that you've broken the bone in the same place, which is rather strange."

"Why's that?" I asked, my voice laden with disappointment.

"Usually when a bone heals, the spot where the break occurs is thicker and stronger. So if the bone breaks again, it's in another place." He brought the x-ray close to me and held it up to the light. "See here, that line right there shows you've gone and fractured the bone in the same spot."

"What does that mean?" I asked.

"Well, that you're going to have to rest longer this time. And I'm going to run a few tests, take some blood and tissue samples, to find out what's causing the problem."

I went home totally discouraged. I was wrapped in a half cast. I could get around pretty easily, but was told to keep all my weight off of my leg for a while. I returned to the doctor after four weeks of rest.

"Good news, Kel. The leg's looking good," he announced.

"Great."

"You probably want to know when you can train," he said.

"You're reading my mind," I admitted.

"Remember those tests I took?" He asked.

"How could I forget?" I laughed. "I felt like a pin cushion with all those needles."

"Well, they didn't show me anything strange. So, I'm going to clear you to train again, but I want you to be careful. I think you should head to the mudflats for any continued training."

"Why the mudflats?" I asked.

"It's where local horse trainers take their race horses recovering from shinny. I think you should run around the bays of the inner harbor. The horses run against the incoming tide, and it strengthens the feeble limbs more quickly because they are forced to pull their legs out of the soggy mud. It strengthens their legs much quicker."

So off to the mud flats I went, and started to train. I quickly found my strength, running there against the tide. When I felt my leg was strong enough, I returned to my normal training regime.

The third time, as they say, is the charm! Early on a Sunday morning before the sun was up, I broke my leg again on a long twenty-three mile run. I had to hobble back to my house, wincing in pain the whole way. I don't know if I was more embarrassed or angry to go back to the doctor.

"Ok Kel, this is getting serious. No more running."

He put my leg in plaster up to the hip and I was completely immobilized for six weeks. I finally relented, and understood that God was answering my half-hearted prayer. He wanted me to put an end to my athletic pursuits.

"God, I hear you. I gave you the option to stop me and you did. Now I'm stopped." I dusted off my Bible and opened up the book I had learned to love and had missed for so long. "I'm sorry for being so thick headed, Lord," I confessed. "I just didn't want to believe that my injuries were signals from you to pursue another path. I really wanted to see how fast I could run and whether I was Olympic material!"

Something in my heart rekindled. I felt that I would, in my lifetime, go before the whole world. It just wouldn't be as a runner for New Zealand.

My destiny as a missionary was excavated from my own dreams. During the time that I was immobile, I dug into God's word. "I wish I had obeyed you from the get-go, God," I prayed before my cast came off. "But I'm thankful for the valuable lesson, and for getting a second chance."

It was very soon after my leg healed for the third time that I quit my job and my athletic training to enroll at Faith Bible College. This was a short-term missions training center in Tauranga. I was starting to realize I had done nothing of any real value or significance with my life. I had just earned money and raced. So I stopped running races and began to run in the path of God's command. It was soon after Bible College that I would once and for all choose the missionary's life, and more importantly, understand that it was this life that God had chosen for me.

Chapter 5: Asian Circle Tour

I enjoyed Bible College, and soon became eager to apply what I was learning about ministry in the mission field. It was 1970, and a lifetime of adventure opened up before us. At the end of our course, a visiting missionary came to our school to introduce us to a missionary organization that sent young people on short-term trips.

"Who here wants to experience God and live out their faith like they've never experienced it before?" The question stopped me in my tracks. A visiting missionary, Ross Tooley, came to Faith Bible College to invite us on a trip with Youth With A Mission (YWAM). "It's a fourteen month trip that will change your life!"

I raised my hand in a flash. I looked around the room. My friend Mike Shelling had his hand up, too. Eventually there were eleven of us from all over New Zealand eager to go.

Everyone assembled in Auckland for a short period of training before flying to Asia for the fourteen month adventure. During this time, Ross and Margaret Tooley, and Barry and Kay Austin, taught us many principles on the "how-to's" of evangelism. Ross invited a guest to teach on the principles of prayer and how an active prayer life was a key to a dynamic life of faith in ministry. Joy Dawson's teaching blew us away. (See Appendix: Principles for Effective Intercession by Joy Dawson).

"How many of you really believe that God answers prayer?" she asked. We all put our hands in the air.

"Great, then we're off to a good start," Joy said. "Prayer is absolutely essential to any Christian worker, either here or on the mission field. I've been invited to teach you some of the exciting principles God has shown me about intercession."

We all eagerly pulled out our notebooks and pens. Joy had a gracefulness about her that put us all at ease, while at the same time she was a "fire-ball" for Jesus, one awesomely enthusiastic woman. "You are all God's children and as part of His family, you can all hear His voice," she told us.

Joy led us in the practice of preparing our hearts to hear God before we prayed by simply worshiping God for His goodness. "It's a privilege to join the Lord Jesus in His wonderful ministry. Praising God as we pray prepares our hearts for prayer. But we don't stop there. We must confess any sin to make sure our hearts are clean before God. We need to be sure to not have any unforgiveness or resentment toward anyone as we pray. Job had to forgive his friends for the wrong things they believed about him before he could effectively pray for them. Don't you think it's the same for us?"

I had never received such a clear teaching on prayer. As I entered into the process that Joy outlined for us, my mind started to clutter with other thoughts.

"Some of you may experience that your own imagination or thoughts get in the way when you pray. You need to learn to die to these things, and acknowledge to God that you can't pray without His direction, or without the energy of the Holy Spirit. Your enemy, the Devil, wants to keep you from hearing God's voice and to discourage you from praying. But he's a liar! You can hear God because you're His child, so thank Him for all that God will show you."

"Lord, show me how to pray." I whispered. "I expect you to direct me because of your love for me."

What I didn't realize at the time was that these principles would become foundational to my whole life of ministry. As students we started to see prayer as an active conversation with God, and that this active conversation was a key to effective ministry. We discovered God could lead us and give us the wisdom and the direction we needed for each circumstance. These principles had been in the Bible all along, but Joy dug them out of scripture and shared them with conviction and power. These scriptures captivated me because I was hungry for truth.

"Have you ever heard someone speak like that on prayer?" Mike asked me after her teaching.

"Never," I admitted. "But if YWAM has teachers like that, then I'm impressed!"

"That's how I felt," Mike said. "She shares the truth she's experienced and lived. I want to do that!"

"Me too," I laughed. "It makes ministry that much more exciting."

No sooner had we gathered for training than we were off for the Asian Circle Tour: a trip to the Philippines, Thailand, Malaysia, Singapore, and Sri Lanka for more than a year. Our purpose was clear. We were to go and preach the gospel in each country before returning to New Zealand.

We left with great enthusiasm and expectation. "I can't even imagine what lies ahead," I confided in Mike as we headed to the airport. "There's so much uncertainty."

"I know," he replied. "I can hardly wait."

I was thankful I would be able to share the adventure with my good friend. The two of us, together with the rest of the team, got on a plane and flew to the Philippines. We were all hoping for an amazing experience, every one of us full of faith that God would use us to share His love. All of us got more than we bargained for: the adventure of a lifetime.

Chapter 6: A Whole New World

The airplane dropped suddenly, and I reached for the seat in front of me. The whole cabin shook.

"The captain has put on the seatbelt sign," the flight attendant's calm voice said over the intercom.

"Another pocket of turbulence Kel. You were asleep for awhile," Mike said to reassure me.

I rubbed my eyes, getting my bearings after being jolted awake. We had been in the air for hours. "How much longer till we land?" I asked.

"They said we'll start our descent very soon. I wish I could have slept like you," Mike said, "but I'm too excited."

We flew into Manila late at night on a lumbering DC-10. Though it was the middle of the night when we stepped off the plane, it was like we stepped into a sauna. The air was thick and muggy—we began to sweat immediately. It was a rude awakening for a farm boy from clean green New Zealand to experience Manila's smog. At the time, I could only guess how heavy and hot the air would be in the middle of the day.

"Welcome to Manila!" A short, smiling man from the local church beamed at us as we got off the plane. He held a sign that said "YWAM" on it. "I probably didn't need the sign," he laughed, "It's not every day a dozen or more Westerners arrive in the middle of the night."

We got our bags and loaded the man's truck with the whole tour's luggage. "A few of you might want to jump in the back to watch the bags," he warned. "It's not uncommon for foreign luggage to be stolen from the back of trucks."

"After you, Mike," I said and jumped up into the truck well, making a seat out of a few of the suitcases. We waved goodbye to the group who went by taxi to the church where we'd stay.

What a way to be introduced to the country I would soon fall in love with! It was a country that would be my home for many years. We got a bird's eye view from the truck as the city revealed itself in its unique variety. Millions of people and vehicles going in all directions, our truck weaved in and out of five lanes of traffic and often the footpath became the sixth lane for impatient drivers. Then there were the incredible smells and sights! Some were pleasant, some were shocking.

The back streets were littered with smoldering rubbish. Rats scurried between the burning heaps, looking for food to carry away into the cracks and holes where they hid. Dogs were everywhere, bone thin, and ravaged with mange.

And the slums! Open sewage wafted into our nostrils, striking us with its putrid stench. Entire villages were comprised of makeshift homes where people lived in squalor. Villagers stood outside small huts made from sheet metal, discarded wood, plastic and chicken wire, and they cooked over open fires.

Manila was chaotic in those days. Everyone, it seemed, drove with one foot on the accelerator, and one hand on the horn. Traffic weaved and throttled at what seemed a dangerous pace, with no apparent rules or order.

"Can you believe what you're seeing?" I shouted to Mike as the truck blitzed through the streets, weaving around potholes and men on little scooters.

Mike gripped the edge of the truck bed with one hand, while his other arm was protecting the bags, trying make sure they didn't spill out on a sharp turn. "There's so much traffic, even in the middle of the night. It's like the city doesn't sleep!"

All these things were so new, things I could never have conceived. I had never seen images of the Philippines. TV was still pretty new at the time, and we had very little programming in New Zealand. Today, you can see the whole world without ever leaving your living room! But on that first trip to Asia, there was no way to know what the world looked like unless you went and saw it for yourself. From the back of that truck, on our first night in Manila, and in days that followed, I looked at a whole new world with surprise.

Adjusting to a New Way of Life

For the first few weeks our team was based in Pasay City, one of the five cities that make up greater Metro Manila. We struggled to adjust to our new environment. Sleep was hard to come by since cock fighting was a very popular sport at the time; Filipinos kept roosters all over the city to compete in the fight-to-the-death matches. As a result, all through the night roosters crowed, one after the other. It was an endless barrage against our jet-lagged team. And if you got over the roosters, there were the dogs to deal with. The mangy street dogs roamed the city looking for food. They barked endlessly, howling and whining all through the night.

Though earplugs might drown out the noise, and heavy sleepers could ignore the wild sounds of the animals, no one could avoid the unbearable heat. There was no fan in the church where we stayed and no air-conditioning. We simply lay in the heat with no relief, thinking we'd never sleep. Every night we would count the hours, or geckoes; and after that didn't work, the number of barking dogs or crowing roosters. We were miserably hot. We sweated on the air mattresses laid out on the floor until we just slipped out of consciousness because of sheer exhaustion.

"I never knew trying to fall asleep could be so exhausting," Mike yawned from his air mattress. Those first nights in the city were so hard for us to endure, humor was our only escape. We men were all staying on the third floor at the back of a church. The air was still, and clung to our skin. We lay on our backs drenched in sweat.

Preparing for the night was always somewhat of an enterprise. We were six grown men in a church classroom. Each of us had four chairs, one for each corner of our air mattress, which we used to tie the strings of our mosquito nets to.

"The key is to drape the mosquito net over your air mattress and tie each corner string to the chairs," our host explained. "Make sure you get the bottom of your net tucked under your mattress—and make sure no pesky mosquitoes are on the inside."

On one of those first nights, as I waited nestled under my netting for the guys to finish getting ready for another sleepless night, I mindlessly started to drum my fingers against the wooden floor.

When everyone was finally settled and lying down, I asked who was the last into bed. One of the guys admitted it was him. "You know, there's an unwritten rule that the last guy to bed has to turn off the light." Reluctantly, the culprit who had forgotten to turn off the light pulled out the netting from his mattress, which he had just painstakingly finished tucking in, walked to the door, shut off the light, and then started the process of tucking the netting back under his bedding in the dark.

I continued to rap my fingers against the floor. “What’s that?” the light-switch friend asked once he was settled.

“Probably a cockroach,” I said jokingly, “and by the sounds of it, it could be headed in your direction.”

There was dead silence. All of a sudden, the light-switch culprit jumped out of his bed, knocking over all four chairs that were holding up his mosquito net, which fell around him causing him to panic as he tried to get free. Scrambling to untangle himself and avoid the bug, he ran to the door, turned on the light and searched around for the make-believe critter.

After minutes of searching, when he couldn’t find the roach, he had to reassemble all the chairs and mosquito net. Finally, he turned off the light, got into bed, and tucked the netting back under his air mattress. I resumed drumming, this time more slowly.

Trying not to laugh when my friend again asked what the sound was, I said, “Well what do you think it is? This one appears to be going a lot slower than the other one, and could be the grand daddy of them all!” Mike and the others struggled to control their snickering. “And believe it or not, it could be headed in your direction.”

After a few moments of uneasy silence, it all became too much for my friend to bear. He leapt up again, thrashing about in the mosquito net, which again toppled the chairs. In his frantic effort to get away from the dreaded invisible monster roach he ended up completely entangled in the netting. After a desperate scramble to free himself, he darted for the light switch again and proceeded to search for the invisible bug.

The rest of us in the room couldn’t contain our laughter any longer. “What are you laughing at?” He asked. He still had no clue!

The practical joke didn’t make the room any cooler, or sleep any easier, but it sure made the hours we waited for sleep more enjoyable. In that time of adjustment, as we became acclimatized to our new environment, and in all those restless, sleepless first nights, I had a lot of time to think.

"God, I'm struggling here. What on earth have I gotten myself into?" I prayed. Everything was so different than what I was used to. The people. The lifestyle. The garbage and slums everywhere! I soon realized that I would be of no use as a missionary unless I genuinely loved the people, but the difficulty of my circumstances made it a challenge to focus on them. "I need Your heart for the people of the Philippines."

And with that confession, an idea came to mind. I started to get up early in the morning to go into the chapel of the church while it was still cool to pray for God to give me the heart I needed, instead of just waiting and hoping that it would somehow come on its own.

Each morning I would wrap my sheet around me because the mosquitoes were so bad, and I would cry out to God to give me a love and a burden for the Filipino people.

Morning after morning, I would go down to pray for God's love until one glorious morning God's compassion was poured out over my entire being. The tides had turned: God answered that cry of my heart. Great waves of God's love and compassion washed over me. It was then that I truly knew God had given me His love and calling for the people of Asia.

Chapter 7: Evangelism 101

I had preached the gospel a grand total of two times in my life before going on that first trip. I was as green as grass! I was hesitant at first to preach, unsure of myself and unfamiliar with any technique. In New Zealand, going door-to-door sharing the gospel had been a bit discouraging; it was most often the case that people were unwelcoming or didn't want to hear what you had to say. So when our team went out for our first day of ministry, a number of us were understandably nervous.

"I don't really know what to say," I confided to Ross. "Doing evangelism in New Zealand and going house to house is one thing, but we're now in a foreign country."

"No worries, friend," he said and patted me on the back. "The same concepts and principles we've been teaching you over the last few weeks apply in most every situation. Just go out and share. You'll find the Filipinos very open to you!"

We headed out to a local park where Filipinos gathered with friends and families. We took a few guitars with us and sat down as a group. Graeme, one of our teammates, started to strum out a worship song while the rest of us sang. People in the park, surprised that a group of foreign students were singing, gathered around us.

"Just go and say hi," Ross said. "Be a friend. Ask them about their lives and share yours."

"That's it?" I asked.

"Yeah," Ross said, and went out to a group of young men that clearly wanted to talk.

"Well," I said to Mike, "here goes nothing."

Ross was right. Evangelism wasn't rocket science. And the Filipinos made me feel like a million dollars! It wasn't uncommon for us to enter a village or basketball court with our message and a handful of tracts, to see those gathered hanging on our every word and then race up to us to grab the tracts. They hungrily read through the small booklets, pointing at the pictures and sharing them with their family and friends. I had done similar outreaches in New Zealand, and we found the tracts we handed out discarded, thrown on the ground or in trash bins.

"They're really interested in what we have to say!" Mike said in surprise in those first days of ministry. "This is a lot more exciting than I'd anticipated."

"I know," I said, glad that Mike felt the same way I did. "They're receiving us outsiders as genuine missionaries. They're not hung up on the fact that we're young. I mean, they expect that we know a lot and want to learn from us!" Our team and materials were warmly received. Without the Filipinos even knowing it, they had provided much needed healing for me just through their willingness to hear the gospel. It had never crossed my mind that such a place could exist, a place where people really wanted the gospel. What a blessed joy to be among such people!

Ross and Margaret had effectively taught us how to teach evangelism, and they set up a tour for our team to visit churches and Bible Schools. We stayed ten days at each location. We'd teach and do evangelism in the morning and the afternoons. Then in the evenings we'd conduct an evangelistic meeting. We didn't require the new believers to go through years of seminary or other training. Instead, we took them with us and had them share their testimonies on the very same day they had received Jesus.

Locals stopped to listen to us, curious why Westerners were out singing and speaking in the area. We would invite them to our next evening service. Many people came to our meetings, got saved, and then went out the next day with us. They were willing to share how Jesus had marvelously changed them, set them free from all manner of vices, and given them hope and a future. The faith we introduced them to was powerful and active. They enthusiastically shared their testimonies of how God had changed their lives, forgiven them and granted them eternal life. They were full of gratitude toward God, and more than happy to join in with us as we shared the gospel.

It wasn't uncommon for us to see God double the size of many churches after only ten short days. We did this from place to place, and church to church.

From Personal Evangelism to Preaching to Whole Groups

After being in Manila as an entire team for two weeks, Ross split our team in half. He took five people and went north. I unexpectedly found myself in charge of the second team. We went south, so we didn't see Ross and the others for about two months or so. Leading was a whole new experience.

I had been doing the brunt of the teaching, but I do recall on one particular morning asking one of my team members if they'd preach. "Would you take a turn today?" I asked, rubbing my hands together in expectation, hoping for a positive response. I wanted a break.

The team member I asked just looked at me and said "No." I asked the other team members, but none of them felt up to it either. What was I to do? People were coming to the services, filling our venues. I had hardly ever preached the gospel before in my life; yet now, here I was preaching night after night. Fortunately, God wasn't limited by my stumbling attempts at preaching. Despite my inexperience, people miraculously continued responding to the altar calls and getting saved.

Overwhelmed, one evening, I approached the interpreter before the service. He was a local evangelist. "Marcello," I said, "could you please be the one to preach tonight? I'm completely out of sermons! I only have four or five sermons in total, and my team is getting really tired of hearing them. By contrast, you're a seasoned evangelist," I reasoned, "if you'll do the preaching tonight, then I could learn from you and my team wouldn't be subjected to another re-run."

He looked at me and smiled. After a few moments he shook his head. "They can hear me anytime," he said. "The people are coming to listen to you foreigners. Sorry brother, but that means you're on."

And that was that. So, before the service I found a quiet place to hide away and cry out to God. "Lord, I need something to share. Please help!"

Many times I'd be at the pulpit, still figuring out my message as I preached! God used those desperate times, even with all of our inexperience, and was very gracious to us. We led many hundreds of people to the Lord, and had the thrill of baptizing many of our own converts in rivers. We were given the chance to disciple them and to teach them to evangelize as a way of life. We would take our converts with us the day after they accepted Christ to share their testimony. It wasn't unusual for converts to help lead others to Christ as "one day old" Christians.

Chapter 8: Good Things Take Time

We were introducing people to God from all different walks of life. Many made commitments to Christ immediately upon hearing the good news of God's love, others took more time. Ross pulled me aside one day: "I want to introduce you to someone, Kel." He stood with a young, and rather gangly, yet good-looking Filipino student who wore designer jeans, and an easy smile. "This is Ray Yap."

We shook hands and talked briefly. Ray was the son of a medical doctor, and a student at Holy Cross University. I could tell from the way he handled himself that he was a natural leader and a popular guy.

Ray was an influencer and had many friends. His likeable leadership qualities would later deeply impact and the local barrgadas, neighborhood gangs comprised of young people who were all up to bad stuff. I was told one of the barrgadas had a knack for kidnapping dogs. They would hit upscale neighborhoods on Saturday nights, sneaking around the back of houses until they spotted their prize: a nice, healthy-looking dog.

They would use a bamboo pole that had a noose hanging from one end. Once they'd chosen their canine victim, they would reach with the pole over the fence, slip the noose around the unsuspecting dog's neck, and then pull the pole until the noose became tight. Before the owners had a chance to know what was going on or why their dog had been barking, the adrenaline filled gang would already be tearing down the alley in their car, whooping and laughing with their catch in tow.

In many parts of the world, a dog is man's best friend and guardian of the family home, a loyal companion. In the Philippines, dogs are all these things and more: they are also delicious delicacies.

"No way!" you might say, "that's grotesque and surely illegal!"

Indeed it was. Nevertheless, once the dog had been dragged to it's death, the gang would celebrate their catch with a barbecue.

"I think it's important for you to get to know Ray," Ross said. "I'd be more than happy to," I said. I committed to spend time with this young man, and told God I would do my best to connect with him every day until he made a decision for Christ. And so for the next several weeks, I'd regularly go over to Ray's house and either invite him to our place or hang out with him and some of his friends.

At the time, our whole YWAM team was renting an apartment above a small storefront church. The church building had originally been converted from a horse barn, so we were actually staying in what was formally the old hayloft of a stable. The corrugated tin roof had no insulation. The tropical sun would beat down on the tin roof all day, making the space utterly intolerable. Somehow, though, we all managed to survive, choosing to make the best of our circumstances.

One day, Ray stopped by our place for an afternoon to visit. He had to ascend up an almost vertical ladder, which was the only entrance. It reminded me of climbing up to a tree house, there were no normal stairs! When he saw where we lived and felt the unbearable heat, he couldn't believe his eyes.

"Who in their right mind would live here?" he asked.

"Probably no one," I laughed. "But it belongs to the church we're helping and was made available for us to stay in. Besides, it's located in your neck of the woods, right where we want to be."

"You're crazy," Ray said. "Let's get out of here. How about I buy you a cold drink to forget about this little oven house of yours?"

Some time later I was given responsibility for the little church and became their pastor. I had known Ray by this time for many weeks, and was constantly meeting with him. We had become close friends. One Sunday morning, as per usual, Ray was sitting in the back of the church, politely listening while I preached. As we neared the end of the service, Ray shot up out of his seat and bolted down the aisle to the front of the church. He surprised all of us.

When he arrived at the front he fell onto his knees. "I have to give my life to Jesus!" He blurted.

I stood there, unable to hide my surprise. I hadn't even given an altar call yet! "Of course, Ray. Let's pray."

Ray prayed and accepted Christ into his heart. Later, when the rest of the congregation had filtered out of the small church and we were alone, I sat with Ray at the front of the sanctuary. He had a smile plastered across his face.

“What was it that caused you to be so convicted in this morning’s service?” I asked. “We’ve spent time with each other and you’ve heard me share the gospel nearly every day. What was it about this morning?”

“All those times of hearing the truth I guess finally became overwhelming,” he said. “I've enjoyed our friendship a lot, listened to you share, and watched your life close up. You’re the real deal, Kel. You have a sparkle in your eye and you live your message. Over these past weeks I’ve become more and more convicted of my selfish and sinful way of life. It’s not religion you and your team are into, it’s reality. I've witnessed the energy of your lives, and I just finally became overwhelmingly convinced of my need for Jesus to become my savior too. I wanted my sin forgiven, and to know for sure I’m going to heaven. I wanted what you have!"

Ray went on to become a true disciple of Christ, and shared the gospel fearlessly. His genuine faith would see many people transformed, including gang members from those local dog-catching barrgadas.

Laughter in the Rain

We discovered that life on the mission field was filled with surprises, and more than a few laughs. We went to a little village on the Island of Mindanao called Wassian, and a monsoon was upon us. Rain fell with violence. It seemed the relentless downpour would wash us out.

We were stuck in the village for four days. We couldn’t do any evangelism because of the deluge. We were all staying in the pastor’s house. It was a small hut built on stilts with a thatched roof, thatched walls, a slatted bamboo floor, and an outhouse that stood apart on the side of a hill. Because of the terrible rain, we all put off going to the bathroom as long as we possibly could.

A portly gentleman had recently joined our team. He loved evangelism and had a vivacious love for God. The time came when he had to go to the bathroom and couldn’t wait any longer. The trip to the outhouse in the rain was an adventure in itself, you had to take an umbrella with you since the facility had no roof, and the rains were so heavy they’d still soak you the moment you stepped out of doors. The clay ground was treacherously slippery, it had so much suction it pulled off your flip-flops. So, either you went bare foot, or you wore shoes or boots. The reddish-brown clay would clump and stick to the bottom of shoes and curl up round the edges. Boots became so heavy with built up clay, you thought you were walking with weights strapped to your legs!

We wished our friend luck and watched as he opened his umbrella and headed up the hill. He slipped and slid all over on the place. The thick clay stuck to the bottom of his shoes as he clumsily tried to get to the outhouse. It wasn't only the mud that made the trip treacherous though. There was a large Banyan tree on the property whose protruding roots covered the ground on the way to the outhouse. Some of the roots grew to a couple of feet high and snaked all over the ground. They were dangerously slippery if stepped on in the rain. Because of this, our friend was forced to crawl on his hands and knees to get over the protruding roots on his way up the side of the hill.

When he finally made it to the outhouse, muddied and wet, we watched him disappear into the small thatched hut. The umbrella popped up above the roofless structure to protect him while he sat and took care of his business. Because the ground was so slick, though, once he was inside the outhouse, he slipped and fell, knocking the whole shed over. We watched in shock as the four flimsy walls of the outhouse crashed over and started sliding down the hill. Only the seat with our friend sitting on it remained, vulnerable and exposed in the rain. Fortunately he had not been injured in any serious way, though his pride certainly took a beating.

We'd witnessed the whole ordeal. As soon as the shock of it all had passed, we began to see the funny side of things. As he finally made his way back to the house, by some great miracle of God, we all managed to compose ourselves. Our friend had not been able to see the funny side of things at all, he was embarrassed and not a happy boy! No one said a word. All his dignity had gone down the hill with the toilet.

When the rains stopped we were sure to make amends with the church, so we replaced the toilet, a small price to pay for the unforgettable memory it brought our team.

Our time in the Philippines soon came to a close. I didn't realize it at the time, but this nation, with it's beautiful people, mountain tribes, rich colors and savory foods, would become the nation where I would spend much of my life in ministry. The Philippines was the place where I found my feet as a minister of the gospel. It was a country I would come to love, lead and minister in, find love, and experience heart-wrenching tragedies. In the meantime, I had other countries to visit with the team. Off we went to Thailand!

Chapter 9: Thailand & Samai

The world in 1970 was certainly a different world from the one described in the stories I read as a child of the young and adventurous missionaries who inspired me. As I traveled by air to Thailand from the Philippines, I couldn't help but think how Hudson Taylor was willing to sail to China by ship. The trip was treacherous, and even by getting on the boat he was risking his life. He barely left the waters of England alive, but God spared him, and the ship traveled on from what seemed like an imminent shipwreck. God gave him safe passage to the nation where he would dedicate his life.

When Hudson Taylor landed in China, he quickly had to learn the Chinese language in order to communicate the message of God. I remember reading as a boy how Taylor was having trouble reaching a group of young Chinese, even though he had become fluent with the language. He was confident his message was powerful, yet each person who had gathered to listen to him was hopelessly distracted. When he finally asked those gathered what they were thinking about while he was preaching, they were embarrassed and didn't want to tell him. They finally confessed they were trying to count the many brass buttons on the jacket he was wearing from London.

The cultures and dress of China and England were so different, Hudson Taylor realized his strange and foreign clothes were a distraction to the Chinese. Soon after that meeting, he changed the way he presented himself. He began to dress and cut his hair like the Chinese. This act of identification cut through the cultural barrier, and his message could be received. He was willing to preach the gospel to the Chinese—he did whatever was necessary to be effective.

Wherever we went on our Asian Circle Tour, we learned about the cultures and tried to adapt to them as best we could. We discovered, though, that we did not face the same cultural barriers that Hudson Taylor did. In fact, people were often drawn to us and gave us their attention because we looked different. Often, it was because locals were curious about the music we sang as we strummed our guitars to draw a crowd. The people were excited and interested to meet young foreigners who had come to their country, so they listened to the Good News we had to share.

Because our trips were short-term trips, we also did not have the time to learn the language in each of the countries we visited like Hudson Taylor did when he went to China. We would learn basic phrases, how to greet people, and relied greatly on translators to share our message of Good News.

One of the Thai translators that partnered with us was a man named Somkiat. He was a university graduate, and was working for a pharmaceutical company when we met. He spoke English very well, and was a great person to work alongside. At first, he was decidedly timid and shy, but he willingly served as my interpreter nonetheless. We were both around 23 years of age at the time, and we became fast friends, walking the streets of Bangkok sharing the gospel. The more Somkiat interpreted, the more he grew in ability and boldness; the Holy Spirit forged us into a strong team.

It's a trend I've seen throughout all my years of ministry. When young people are given a chance to minister and risk sharing their faith despite their shyness or fear, God works powerfully through them. This was as true for Somkiat as it was for me. I found that I was becoming increasingly confident in the message I was sharing. I saw the power of God change people's lives as I stepped out in faith and watched God work in people's lives.

On our Asian Circle Tour we were leading people to Christ virtually every day. Many times we were privileged to baptize our own converts, dedicate babies in local church services, serve communion, pray for the sick, and were blessed with opportunities to disciple and teach our converts. The field training I received by stepping out in faith to minister was powerful and effective; it helped to prepare me for the exciting years that were to follow.

When Somkiat had finished his work for the day, we would hit the streets of Bangkok, going door to door at high-rise apartments and talking with people about Christ. The occupants in the apartments were predominantly Buddhists. We had wonderful success going door to door and won many of them to Christ.

We would work late into the night, so I would often get in trouble with my YWAM team for getting home late. There was a six-foot concrete wall surrounding the whole compound with shards of broken glass set into the cement on the top of the wall to discourage people from climbing over it. At the entrance was an equally imposing spiked steel gate which could only be accessed from inside. I was in the habit of getting locked out and having to wake up a teammate in the middle of the night to let me in, so I didn't have to sleep outside.

One night I said goodbye to my friend Somkiat, and hurried home, trying to avoid waking up one of my teammates. The road we walked on to get to the YWAM compound passed a series of jewelry shops. Every night, I noticed the same guard working the late shift, a large automatic rifle resting on his lap. He guarded four jewelry shops all through the night into the wee hours of the morning.

As I walked by the shops, thinking about the evening I had spent sharing the gospel in downtown Bangkok, I noticed the guard and felt an inner prompting to witness to the man. I quickly blew the idea out of my mind because Somkiat wasn't with me. I couldn't speak Thai, and I was sure the guard couldn't speak English. However, the large steel gate—which would undoubtedly be shut if I didn't hurry home—loomed larger than the language barrier. So on I sped to the base.

When I arrived at our compound, I soon discovered that I was locked out anyway. I muttered a prayer under my breath, and was just about to ring the bell when I felt a strong rebuke from the Lord: Go back and witness to that young man, even though you can't speak Thai, and he can't speak English. I turned away from the doorbell and I put my hand down at my side. I would have to wake my friend later, much later. Reluctantly I turned around and headed back down the street toward the jewelry shop.

As I walked along Sukumvit Road, I practiced the few phrases in Thai that I knew. Perplexed and unsure about how I was to share my faith with someone who couldn't speak my language, I walked up the steps to where the guard sat. He smiled at me, while holding his automatic rifle. He acknowledged my approach with a nod of his head and greeted me. After returning his greeting, I was quickly running out of Thai! We looked at each other for a moment, and then I extended out my arm and pointed to the watch on my wrist.

With a series of hand gestures and rather loud, slowly spoken English, I tried to convey to him that I would come back to the jewelry shop the next evening between six and seven o'clock, and take him with me to the missionary compound where our team was staying. My idea was that we'd have a barbecue, and take him swimming in the pool at our compound, gesturing as best I could to explain my plan. When all the communicating was done, we nodded to each other goodbye, and I turned around, heading back down the stairs and along Sukumvit Road toward the YWAM mission base.

I had no way of knowing how much of my little charade the man understood. I was right that he spoke no English, and was also right that I'd have to wake one of my teammates! Mike graciously let me in the compound. I drifted toward sleep, wondering if the young Thai would be there the next day.

He wasn't. A different man sat with a gun lying across his lap. Disappointed, I turned to Somkiat who had accompanied me to the shop. Somkiat was just asking the guard where the man from the night before was when he appeared from an alleyway alongside the shop with a big smile on his face. He held out a brown paper bag that he opened for me to peer into, and there in the bottom was a pair of shorts he'd brought along with him to go swimming. I was so excited that he'd understood our communication from the night before, thank you Jesus. We both had big grins across our faces and happily shook hands. I repeated, "My name is Kel." He pointed to his chest and said, "Samai."

I'd gained a new friend, though all I knew about him was his name.

We had a great evening together. Through Somkiat's interpretation, I learned that Samai somehow had understood everything I had tried to communicate the previous evening, even though he didn't understand a word of English. God had made sense of our awkward conversation. Samai explained that the man who had taken his place-his father-was a devout Muslim.

Samai had never gone swimming before, so it was fun to watch him splash about in the pool. After eating together, Somkiat and I started to share our faith with him. We spoke late into the night about the story of God's people, about Jesus and the work of the cross, and in the early hours of the morning, Samai surrendered his life to Christ.

We discovered soon after, however, that more so than the steel gate locking our compound, and more than our language disparity, Samai's Muslim background was a difficult barrier to break. It was more difficult, at least, than

I had initially understood it to be. For the first week after he'd accepted Christ, Samai perceived himself to be Muslim and Christian. He thought he could be both.

This misconception slowly changed, and it was God's gentle and precise work that broke through the barrier in Samai's life to help him fully embrace Christ. Each day, Samai would walk to our compound and we would teach him English through the stories and language of the Bible. The journey to our place took him the better part of the morning, as his home was quite far from our compound. One morning he arrived with a big smile splashed across his face. He told us that on the walk, tired and bored, he had prayed, and said within himself, "I just wish I had some money for the bus because it would save me a lot of time."

He had barely got the words out of his mouth, when a complete stranger walking toward him on the crowded Bangkok street extended his hand out and shoved a white envelope into Samai's hand. Without saying a word, the man had continued walking. Samai was shocked, who in the world was this man thrusting an envelop into his unsuspecting hand? He was a complete stranger. A bit stunned, soon after the man disappeared from sight, he opened the envelope and found money inside. It was enough for a week of bus fares. Samai's delight and surprise that prayer could be answered, and answered so immediately, was written all over his face as he told us the story. Reviewing the situation, I've often wondered to myself if the mysterious deliveryman that day was none other than a messenger angel sent by God.

He came by bus to our compound each day that week for English lessons using the Bible. By the following week, though, all his money had been spent. After walking many blocks along the crowded and hot Bangkok side walks, he once again found himself praying, "I wish I didn't have to walk. I wish I had enough money to take the bus to save me time."

He happened to be passing the home of one of his uncles that his family saw little of, as he uttered this second prayer. His uncle spotted Samai and called out and asked what he was up to. Avoiding the issue of his conversion to Christ, Samai explained that he was learning English, and was on the way to meet his tutor. Happy to hear of his nephew's education, he asked Samai to wait a minute as he disappeared inside his house. He soon reappeared and came out to the side walk and handed Samai another white envelope. How incredible was this, another white envelope! Samai had never received anything from this uncle before or since. Inside was money that once again arrived right after praying for it. Again, Samai had enough money to take the bus for another full week.

Soon after these experiences of answered prayers, Samai gave his life fully to Christ. "The Christian God is real!" He proclaimed to us. "He really answers prayer."

Samai soon went home to tell his father about his newfound faith. His father's reaction could not have been more different from our joy at the exciting news. The man was furious at Samai's conversion and said to his son, "You've shamed me and the whole family by turning from Islam to Christianity!" He began to chastise his son, and the whole family got in on the act.

Samai became fearful for his life, and ran out of the house. He had only just gotten out of bed, so he ran in terror for his life through the streets of Bangkok in only his pajamas. That's how he arrived to us, tears streaming down his face in our compound the morning after his happy and full conversion.

He was frightened as he faced the incredible cost and loss so many believers must face when they choose to love and serve God. When he was able to communicate what had taken place, and as he considered his future and what the decision for Christ really meant, Samai said words that I will never forget, "I could never go through this if I didn't know that Jesus loves me more than I love my father."

After Samai was forced to flee from his father, he stayed with us at the compound for the time we had remaining in Thailand. We continued to teach him English from the Good News Bible.

My time in Thailand was coming to a close, and I was preparing to travel on to Malaysia and the rest of the Asian Circle Tour. When I left, Samai was working a job outside of the city. When I returned to Bangkok later on in the tour, I asked Somkiat where Samai was and what he was up to. When we learned he was working out at a quarry, Somkiat and I went to visit him. When we found him, he was in good spirits, but it was apparent he missed our close-knit Christian community. He had no fellowship out where he worked, no church, and no way to grow further in his faith. So we encouraged him to leave the quarry, and come back to stay with us.

It wasn't long before I had to leave the country again, however, so we looked for a place where Samai could live and work alongside other believers. We arranged for a Norwegian missionary, who managed a Christian bookstore in Bangkok, to look after Samai. Our team often came to the bookstore to get materials, tracts, and books for our work. We got to know Anna quite well. She was faithful and kind—we knew Samai would benefit under her care. When we left him in her charge, she didn't know what to do with this young Thai convert, so she enrolled him in a three month Bible school. While studying there, a beautiful young lady caught his eye!

The next time I visited Bangkok, Samai had enrolled in a four-year Bible College in the south. I phoned the Bible College to find out how he was doing. I listened happily as the college staff informed me that he had turned the whole Bible College upside down. He had become a powerful evangelist, hitting the streets to preach every time there was a break in classes. At lunch, he would quickly grab something to eat and then go to share the Good News. The College leadership reported they had never had a student like Samai before. He was so eager and active in ministry, and winning so many people to Christ, he'd started a church right there on the street!

Samai's Thriving Ministry Today

On another visit back to Thailand, I met up with Samai, who had just completed two years of compulsory military service. As he talked about his future plans, he said that he wanted to go to one of the most difficult places in the country where Communist and Islamic insurgents were fighting government forces. And that's what he did. With his beautiful and talented wife, Pontip, whom he had fallen in love with during Bible school, Samai decided to journey to the southern province of Surat Thani. The gospel had never before been preached in this place of great darkness where Buddhism and Islam held sway over people, unchallenged for centuries. There were no churches, then, in that region. But Samai labored to change that.

I had the privilege of seeing the fruit of this couple's faithful work years later. Samai and Pontip had planted over two hundred churches. The couple also developed a beautiful church camp facility and a summer Bible School.

"I have something I would love to show you," he said when I met with him years later. At this time they had three grown children, all of them walking with God. We traveled to a riverside where he showed me the property his mother-in-law had donated to the ministry for a church camp and retreat center. The river carved its way around the camp and through lush, green landscape. Samai turned to me, a man much changed from the one I met all those years ago in Bangkok.

Looking at him, I could hardly remember the shy seventeen year-old I had met more than thirty years earlier. Back then, he had a job that hardly paid anything and his future didn't seem promising. However Samai's life was forever changed when he said, Yes to God. God pulled him into a great adventure and persuaded Samai to give his life to the work of the gospel. From a young man guarding a jewelry shop, he had now become a mighty general in the army of God.

Though over thirty years of ministry and life had been etched into the contours of his face, there remained that same joy that had lit up his being the day he accepted Christ. "I would like to take you swimming," he said, smiling. "I owe you for that swim in Bangkok."

When we went down to the river that day, it was not as two young strangers separated by barriers of language and faith, but as brothers, grown up in Christ. We were whooping and laughing in the knowledge that we were both washed and cleansed by the grace of God. It was wonderful playing in the water like we had done all those years ago when we first met.

Chapter 10: Off to Malaysia!

Already our Asian Circle Tour had been underway for nine months! It was hard to believe the time had passed so quickly. So much had happened that it seemed like a lifetime ago that I was at home in New Zealand. It was now August, and in a few short months, this whirlwind evangelism tour and trip of a lifetime would be over.

We arrived in Malaysia by way of the small border city of Hat Yai, Thailand. My travel companions were Maria van Klink and Val Batup, both Kiwis. Other members of the tour followed the promptings of the Holy Spirit to go to other parts of Malaysia where opportunities abounded. This was the first time any of us had been to Malaysia.

A pastor who lived in a city that was then called Port Swettenham had invited our small team to stay. It was a vibrant city on the southern west coast of the nation, and also the largest and busiest port in the country. Situated on the Straits of Malacca, a critical passageway that connects the Indian Ocean to the Pacific, the port is nowadays known as Port Klang.

As we arrived in Port Swettenham, it was not difficult to imagine the stories of great men and women who had gone before us. As I looked at the water, and watched ships come in to port, I could imagine the apostle Paul and his traveling companions. I imagined them embarking on adventure, sailing by ship all those years ago in the first century, as they traveled the Mediterranean to different ports in the Middle East and Europe, arriving in places where the harvest was ripe with people who did not know God.

I've often been amazed in my life that I too am caught up in God's great adventure. What a miracle to have been invited like so many others to go and spread the Good News of His love! With each ship that came or went at Port Swettenham, I could imagine another man or woman of God leaving their country or city to travel to another. Women, like Lottie Moon, had left everything behind in the United States to sail and minister across the world. This single woman left the comfort of home to minister in foreign China for her whole life. I could imagine CT Studd getting on a ship and sailing to China, then

India, then Africa, and his daughter, Pauline, and her husband following in his footsteps to work with him in Africa.

How inspiring to think that I was part of this amazing tradition of men and women who had, themselves, been transformed by the love of Christ! As part of that same mission, which had started in the first century and which is ongoing today, I was setting out like so many others before me, to another part of God's world to share that same transforming love.

We would soon learn that Malaysians are a friendly and hospitable people. This was a quality we experienced throughout our time in Asia, in fact, all along the tour. Ross and Margaret planned our time in Malaysia to be like our time in the Philippines and Thailand, we would go into a city or area for ten days and base at a church. Throughout the week, we trained people, and took the new converts into their own city to practice what we'd taught them. After the whirlwind and excitement of those ten days' work, plus a short break, we'd travel on to our next location to start the process all over again.

We arrived in the first church that had invited us to do evangelism only to discover that the church didn't really exist. We met with the pastor and his wife in the small storefront shop where they gathered with their followers: three young people who weren't yet believers. On our first night with this humble group, we had an evangelistic meeting, and all three of the young people got saved.

We were really blessed as English speakers to realize that most all of the people we encountered in the area could speak relatively good English. We didn't have to rely very much on translators. The British had significant influence in Malaysia from the late 1700's until 1957 when Malaysia gained its independence. As the British Empire spread through trade routes, so did the English language. Communicating our message was therefore never a problem, and we felt mostly free to directly speak with the people we met.

Our arrival happened to be right in the middle of the national school holidays. There were lots of high school and university students out and about, filling parks and other public places. We would go and introduce ourselves to young people wherever they were gathered, strike up conversations, and make new friends. In the process, many of these new friends accepted Christ.

The following day, we trained them, and took them onto streets and out to parks and encouraged them to share their new faith in God. Each day, we had more people in training. Consequently, we were able to do more evangelizing. By the end of the ten days, we had about seventy converts! Each day, we all crowded inside a little shop front in Port Swettenham. We shared our faith in God, the principles we had learned about prayer, and godly living and we taught many stories from the Bible.

Val, Maria, and I watched in astonishment: every day young people were coming to the shop in greater numbers.

"The young people here are so hungry for the truth, they'll risk their lives to get here!" Val said before our meeting.

One of the young men had a small motorbike that he would pile up with friends so that they could come to our meetings. To get all the young people to church, he had somehow managed to load up his motorbike with eight people at a time to save them from having to walk. I was always amazed at the extraordinary sight of eight young adults—somehow standing, sitting, crouching and huddling on a tiny teetering motorbike. How the driver managed to balance the bike, piled high with eight passengers, down the road I'll never know. A small miracle in itself that they never crashed!

We had a great time, the young people didn't want us to leave. God was with us. We had baptized many of them ourselves, and became close to them in those ten days. We had a schedule to keep, though, so we had to go our way. We traveled to our next ministry location, which was a small town, a little further south. Here, we were received and blessed just as we had been in Port Swettenham.

At the end of our ten days, we had about sixty more converts. The young people in this town proved just as responsive to the gospel as the young people had been in Port Swettenham. They loved us, and were eager to become soul winners themselves. Ten days after our arrival, they were already leading their friends to Christ!

A young woman and her friends ran up to Val and hugged her as we said goodbye. "We don't want you to leave!" As we left, the others swarmed us and asked if they could have our addresses because they would miss us, and wanted to keep in touch. We started to write down our contact information on the small pieces of paper our friends were handing us, but I soon realized that it would take a long time to write out the information for them all.

"Friends, we're going to miss our bus," I said, mindful that the clock was ticking. Not wanting to miss our bus, and not wanting to write out my address fifty times, I went over to the chalkboard in the classroom where we'd gathered in the church. "How about we simplify things. I'll write down the address here, and you can all record it on your own." We said our goodbyes, left our friends, and stepped into our next part of the adventure.

The following week, there was an annual conference being held for that particular denomination in Malaysia. As the leaders of the church gathered, they asked the pastor, who had hosted us at that second location, for a report about YWAM. The pastor gave a negative report, saying that YWAM was trying to steal its young people from the church.

I discovered then, as I would discover later in ministry, that the enemy of our souls uses many tactics to cause division and sabotage to the work of the Kingdom of God. We only meant and desired to keep in touch with our friends through letters. But, jealous of our influence, the pastor explained how the address we left on the chalkboard was proof of our plan to ensure that the church's young people follow us rather than the church.

By the time we arrived at the next location, the damage was already done. Because of the bad report from that one pastor, all the other engagements we had set up in churches throughout Malaysia with that denomination were cancelled. Suddenly, we had no schedule. That was a huge blow to our team. We had previously been invigorated with faith and momentum. It felt like God's plan for our time had been totally derailed and stolen out of our hands. Talk about discouragement!

Because everything was cancelled, the pastor in Port Swettenham invited us to come back to his church. Confused and disappointed, we arrived back in the busy, bustling city. We were refreshed to see our friends. We were also thankful that the power of God had continued to reap a great harvest. After ten more days in Port Swettenham, we had grown to around one hundred and fifty converts! A church was born.

The Gospel for All People

After spending the mornings in training, some afternoons we would go door to door through neighborhoods and share our faith with strangers. One such day, the young man I was with stopped me as we walked up to a certain residence.

"Brother Kel," he said, stopping me in my tracks, "we can't go here."

"Well, why not?" I asked.

"I know who the people are who live here. The man works for the police!"

"Oh," I said, not yet understanding why this would be a problem. "Well, don't policemen need to know Christ too?" I asked.

"But in Malaysia many police, like this man, are Muslim."

I suddenly understood the young man's hesitation. This was a common hesitation that many Christians have when they consider sharing their relationship with Christ to people of other faiths, especially within Muslim nations like Malaysia where it is illegal to convert a Muslim to another religion. The temptation is to avoid sharing the gospel to any Muslim in Malaysia for fear of going to jail.

I tried to assure my nervous friend that it would be okay, so I walked up to the door and knocked. Sure enough, the policeman came to the door, and we started to talk. We told him about ourselves, and had a pleasant conversation. He asked us pointed questions, which gave us the opportunity to share our faith. We laughed together as we became friends. The man did not accept Christ, but he did listen to us and was very open to hear what we had to say. He was a kind and pleasant man.

When we left his house, I turned to my friend and asked, "Now that wasn't so bad, was it?"

My friend agreed.

"That man isn't a Muslim because he was converted into Islam," I said, "he's a Muslim because he was born into a Muslim family. Before today, he had never been able to even dialogue or think about the difference between Islam and Christianity, much less make a decision for himself between the two."

"But he didn't choose Christ," my friend said, confused.

"No, but good seed was sown. You and I can water that seed through prayer and the Holy Spirit can cause it to grow, creating in him a hunger for truth. He didn't convert but he's more informed than he was. He won't be able to say to God on judgment day, 'Because I was a policeman and a Muslim, Christians gave up on me and didn't preach the truth.'"

"But what if we had been arrested for sharing our faith?" my friend asked.

I looked at my friend with a smile. "Then we'd get arrested. We shouldn't refrain from sharing our faith because of man-made laws. Today, did we allow the fear of men or their positions of authority keep us from sharing the gospel?"

My friend stood and thought about this for a moment. "Brother Kel?" he asked, "Are there any times when we need to be careful about whom we share with or how we share our faith?"

"Of course," I said, thankful for his question. "We have to be prayerful, wise, and loving and must not be pushy in the way we share. We ought to allow the Holy Spirit to guide conversations in such a way that we provoke questions in the hearts of the people that are both worthy and deserving of answers."

"How do you do that?"

"Just like we're talking now," I explained, sitting down on a bench to rest my tired feet. "I have conversations with people because I'm interested in them, as a friend. Extending love and friendship to people has a positive disarming effect, and gives people the confidence to ask questions, it also gives me the right to answer those questions. Every experience we have with people is an opportunity to shine light and truth into darkness."

Just as I encouraged my young friend that afternoon, I continue to encourage young people all over the world today to make an effort to talk with people wherever they go. I've found that the best way to connect with people and get to a point where you can talk about God is to be interested in them and their lives. When I talk about God, I have fun and keep things light. Evangelism doesn't have to be a scary or heavy thing, even with people who we might fear to talk to, like that Malaysian policeman. God is the one who works in people's hearts. It's our job to knock on the door and risk making an introduction.

A Lesson about Anger

We were in Malaysia for more than a month. Because of our unexpected change in schedule, we spent much of that time in Port Swettenham. And although it was a rich and blessed time, it was not without frustration. The more I kept thinking about that pastor's negative report at the conference, the more irritated I became with him. Because of his bad report, the work of God had clearly been impeded.

His misrepresentation of YWAM bothered me, but what really got to me was thinking about how all the other churches were now missing out on the work God wanted to do amongst them. I really wanted the pastor to know how I felt. In all honesty, I wanted to get on the phone and give him a piece of my mind. However, every time I went to call him, I felt unrest in my spirit. Each time I went to make the phone call, a question would form in my mind. Are you still angry Kel? I was! Then another question would rise up in my heart: Can you forgive this man?

I told God that I could forgive him, but still felt frustrated and angered by the actions of this pastor. I just knew, somehow, that I was not to phone him until my heart was at peace. So I gave up on phoning him, and decided instead to write him a letter. Still, every time I went to pick up a pen, resentment burned on the inside. Again, the Holy Spirit asked, Are you still angry Kel? Each time, I had to admit, yes, and I would put down the pen.

It took many weeks before my heart was sufficiently at peace so that I could write an acceptable letter—one that came from a forgiving heart that wasn't full of my own anger. I'm so thankful I waited until my emotion subsided. Through that experience, God reinforced in me the truth that He, not I, is judge of the whole earth.

We were gravely misjudged and wronged by the pastor, but even though his actions hurt the Kingdom of God unjustly, I had to recognize that I was not the judge and it wasn't my place to set him right. That was God's territory. Until I could communicate straight facts for the record, without injecting my emotion into it, I had to leave the matter alone. That pastor was a Son of God, just as I was. My place was to let go of my own hurt, and commit it all into God's hands.

Learning not to respond in anger was humbling. I might not have been able to change the circumstances I found myself in, but I could control my behavior, I could stop myself from making the situation worse. Eventually I was able to write a letter that was kind and fair, one that addressed my concerns. There was no trace of anger or judgment in it. I never did hear back from him, so I left the matter in the God's hands.

Years later, I heard Billy Graham put into words the reality of my experience when he was asked by a well-known TV presenter in an interview why he never criticized other people. The world famous preacher thought for a moment and then replied, "It's the Father's job to judge, the Holy Spirit's job to convict and my job to love." What a shining example those words were to me.

Thankfully, despite that discouragement in Malaysia, there was much to be joyful about. The good seed that we sowed during our time there has multiplied. To this day we continue to reap a bountiful harvest of souls there. God graciously honored our work. At some point along the way, the pastor in Port Swettenham felt that he had done what God had called him to do in the church, and handed leadership of it over to Pastor Henry Rumiah. Since assuming leadership of the church in what is now Port Klang, the fellowship has grown like crazy!

As it stands today, the church has become the biggest in Malaysia: it has more than 10,000 members and has outreaches in countless countries all over the world. Like the port city where it resides, the church is an international hub, a port and connection between Malaysia and the work of God all over the world. We had good seed that went into good soil, and it was properly tended by those who remained to lead the church. The fruit is forthcoming to this day.

Chapter 11: The Miracle of Provision

By the end of the Asian Circle Tour, our whirlwind fourteen-month introduction to missions, Mike Shelling and I returned to the Philippines. We both felt called to actively minister in this country for the long term. In the Filipino capital of Manila I led a small YWAM team to a fishing village called Cavite. It was a few hours outside of Manila. I went with Mike and a few Filipino believers. We were doing evangelism in the fishing village and saw many people get saved. Before we completed our outreach, however, we ran out of money.

We didn't have a dime between us, and we still had a few days of planned meetings left in the village. I remember getting up early this particular morning to pray and seek God. I was letting Him know that we were fresh out of money, and that we would not be eating breakfast (or lunch, or supper!) unless He somehow provided for us.

After putting all this before the Lord before the rest of the team was yet awake, I decided to leave the thatched hut where we were staying to talk with the Lord out loud.

When I went to open the door it appeared locked, which was rather strange considering I had just finished pulling back the sliding bolt that secured the door. It wouldn't move. It had rained heavily the night before, so I assumed the wood had swelled and jammed. The whole team was asleep in the hut, tired from yesterday's intense ministry. I had a real problem on my hands. I didn't want to wake the team, but there was no other way outside except through that door. So there I was, struggling to get out the front door, and trying not to make any noise as I fought to get it open. After failing at the "being perfectly quiet" bit, I managed to crack it open a little, enough to peek outside.

A thin line of yellow light spilled into the room from the outside. To my surprise I saw that a large box was pressed up against the door. If I wanted to get outside, I'd have to shove the door with enough strength to push back the box that was holding it shut. So I crouched down and braced myself for a final push. By using my shoulder for more power—I'd altogether given up on being quiet—I finally got the door open.

To my incredible surprise, on top of the box was a very large fish. It was still alive, flailing its tail and mouthing for air! Whoever put the box on our doorstep must have done so only moments before. The box was big enough to hold about forty pounds. On closer inspection, I discovered that the box was full of food. There was a supply of rice, noodles, canned goods, vegetables, and all kinds of fruit. I was flabbergasted. I had barely finished talking to God about this forced fast that was about to be upon us, when here was a wonderful provision at our front door!

The mysterious provision was an immediate answer to prayer. It was a thrill for me as a young leader to see how intimately aware God was of our needs, that he was already moving to supply our present need before I'd even had time to pray!

That morning we cooked that big fish and had a wonderful breakfast. We feasted like kings despite the state of destitution we had been in only moments prior. "This is the best meal we've had in weeks," Mike said as we ate with gusto. That breakfast was a feast of happiness. In light of God's provision, it was also one of the tastiest meals in my life.

Learning to Trust God to Provide for Me

I love the old stories in the Bible of how God cares for his people: how He fed the Israelites with manna all those years in the desert of Sinai, how He provided for the needs of David and his men who lived all those years hiding from King Saul, and how He fed Elijah through meat-carrying ravens in the desert during Israel's terrible three-year drought. God even provided a coin for Peter in the mouth of a fish when he needed the money to pay his taxes.

I'd always known and believed that God was a miraculous provider through these stories. However, during those first years in Asia, I had a strong desire to know for myself that God would always provide. I wanted more than historical understanding. I wanted to see that God was the God of the now, who provided for me—the God who takes care of Kel Steiner.

When I first left New Zealand to embark on the Asian Circle Tour, I left with a lot of cash. I had worked hard to save money for that kiwi fruit farm I wanted to buy. Throughout the tour, at different times, I found myself saying to the Lord in my prayer times that I was looking forward to the day when all the money was gone so that it would finally be just God and I.

It was an exciting prayer to pray! Perhaps a crazy prayer for some. But I prayed that prayer, and got used to investing my money in people of the Philippines. I purposefully gave until I got rid of all the money I had saved up while working in New Zealand. There were so many opportunities to give! Pastors in the rural countryside had many needs, even the most basic things like food and clothing were in short supply. Often children had no shirts or dresses, and sometimes no pants. I vividly remember a pastor kneeling at the altar in front of his church, praying. The soles of his shoes had big holes in them, and I could see that he had no socks. Pastors would cycle to outstation churches they'd planted, leaving their families for days at a time to make the trip. A motorbike could cut travel time down to a day. So I bought one for them. It was a joy and privilege to invest in God's people in such ways.

When I learned that Loren Cunningham, the founder of YWAM, was running a six-week Discipleship Training School in Switzerland for young leaders, my attention had been caught. I immediately prayed about whether I should participate in the school. I had an overwhelming sense in my heart that I should go. Excited about the thought of a new adventure, I was rearing to go. I pulled out my wallet to see if I had the funds. What I found was that I had spent virtually all my money. My empty wallet stared back at me as if to say, “Sorry Kel, you're too late. You should have learned about this school sooner, while you were still solvent!”

When the reality of my situation finally took hold, I quickly realized this was an opportunity for God to prove Himself as my provider. This was the very opportunity I'd been praying for. It was finally just Him and me, my wallet had at last been taken out of the equation. So, I committed to going to the school, even though I didn't have the money. I had cash enough to buy a ticket to Thailand, which was an inexpensive short trip from Manila. So I wrote my friend Colin Shaw, in Bangkok, to tell him I was coming for a visit. Colin had gone through Faith Bible College as a student one year ahead of me.

When Colin met me at the Bangkok airport, he advised that I should verify the details of my ongoing flight out of Thailand while we were there at the airport. It was the rainy season, so flight schedules were notorious for changing. The strong monsoon rains also often knocked out phone lines, making things like confirming flights difficult to do. I was embarrassed to have to tell him that I didn't have an ongoing flight.

"Well you'd better buy one now while we're here," Colin said.

"I don't have any money," I replied.

"You don't have any money!" he exclaimed. "Well, then what are you going to do?"

I cupped my hands together under my chin and smiled, signaling that I would pray.

"It's like that is it?" he asked and smiled back. "Well then, let's go to my house. We better start to pray for a way to get you to Switzerland."

And pray we did! The next morning Colin came out of his room with a square tin the size of a small cookie container and put it down on the table.

"What's that?" I asked.

"Maybe it's the answer to your prayers," he said. "You know Kel, for the last six months God has prompted me to put money aside every month. I didn't know why I was doing it, but I did it out of obedience. Last night as I prayed about your situation, I felt the Holy Spirit tell me 'The money in the tin is for Kel.'"

"How much is in there?" I asked.

"I have no idea."

"Well, dump it out, man!" I said, excited about what we would find.

Colin twisted the lid, tipped the contents out, and we excitedly got into the business of counting the money. It was great to see all the coins and bills, but it clearly wasn't enough to make it all the way to Switzerland.

"You know Kel," Colin said after we had counted the money and brainstormed about what to do with it, "there's a charter flight that leaves here early every Wednesday morning. It's a crazy flight, but it goes to London. I think there's enough money to get you there. What do you think?"

London was closer to Switzerland than Bangkok was, so I agreed that we should look into it. We went down to the charter service and sure enough I had enough money for the flight. I had seventeen shillings and sixpence left over

(About $2.50 USD). I later learned that the charter service went bankrupt three months later!

Half Way There

I arrived in London in the rain, and spent about eight soggy days in England. YWAM had a rented apartment in downtown London where I stayed. I had no plan but to get to Switzerland, so in faith I waited to see God provide the next leg of the journey. I kept my eyes open for His provision, and jumped into the work and ministry in which the YWAMers in London were involved.

Each day, we'd go out to Piccadilly Circus to share the gospel with hippies who were gathered from all over the world. The crowded streets of London teamed with life as people emerged from underground tunnels, taking the Tube from place to place. Despite the nasty weather, the drizzling rain didn't keep people from gathering in the famous area to view sights like Big Ben, or the changing of the guard at Buckingham Palace. Tourists who were snapping photographs of the sites constantly surrounded us. I would talk with the young, freedom-loving hippies who gathered with their guitars and talk of peace in Trafalgar Square.

What a thrill for me to be in the nation that had produced so many of the missionaries I admired. Men like David Livingstone, Hudson Taylor and CT Studd. It was almost surreal to share the gospel like these great men, in the country that once sent them into other nations.

After each day of sharing my faith in busy London, I would return to the humble base of operation where I slept. The room I stayed in was very tiny, like a closet. The space where I slept was enough to fit a single bed and a small dresser and that was it! There was room enough to stand beside the bed, and a small area at the foot of the bed where I could hang clothes. It was like a closet within a closet!

Each day in London I was one day closer to the start of the school. I was beginning to really wonder how God would make a way for me to get to Switzerland. On my tenth night in London, at about 10:30 PM after a quiet time with God, I put my Bible beside my bed and started to drift off to sleep. And as I did, the Holy Spirit seemed to whisper to me the encouraging words, Kel, tomorrow you go, son.

I awoke early the next morning to a cold, wet, dark and miserable day; but I was excited to discover how God had provided. As soon as I was up, the first thing on my mind was the money. I was convinced God had told me, Today you will go, so I began to search. Surely the money was somewhere in the room. I looked under the pillow, it wasn't there. I looked all around the floor, I peered under the bed, but I found nothing.

I was fast running out of places to look in that little room. I opened the drawers of the tiny dresser, but, sadly, there was no money to be found there either. Growing more discouraged, I pulled open the small closet door where I hung my shirts. You'll never guess what was in there: nothing! At this point I had exhausted my search. There was no money and nowhere else to look.

I was despondent. "Lord, I was so confident last night that you spoke to me and told me today I would go," I said. "But you need to show me where the money is!"

Then a verse from Matthew 6 that I learned years ago in Sunday school came to my rescue. "Seek first the kingdom of God and all his righteousness and all these things will be added to you." Things like money for the trip to Switzerland. I had been more worried about finding money than with where it came from, I had neglected to seek God first. Beginning to feel convicted about putting my travel concerns ahead of my relationship with God, I quickly asked forgiveness from the Lord and decided to have a quiet time right then and there. I could think about the trip later.

I picked up my Bible, and as I was about to open it, I noticed a book-marker at the front of my Bible, in Genesis.

"That's funny," I thought. "I haven't been reading from Genesis." I flipped the Good Book open to the marker, and there was a five-pound note. This wasn't enough to get me to Switzerland, but it sure was quite a surprise. "I need a lot more than five pounds to get to Switzerland," I said to myself, when I noticed that there was another marker at the end of the Bible in the book of Revelation. Could it be? I opened the Bible to the spot of the marker, and there was indeed a revelation! A ten-pound note!

I then noticed that markers were all throughout my Bible. I excitedly grabbed it by the back cover and started to shake it so all the pages fluttered loose. Suddenly all sorts of bills started falling out of the pages: crisp, clean,

British bills! It was truly an awesome miracle of God's provision. I hurriedly scooped all the bills from the top of my bed and crammed them into my jean pockets.

Once I picked up the last bill, I ran out the door, down the small winding staircase of the apartment and out onto the street. I knew where I was going. I had spied out a travel agency earlier in the week: ticket time! It was early in the morning, so the small travel office wasn't open yet but I could see a woman toward the back of the shop, talking on the phone.

I rapped the glass door with the palm of my hands and peered in, doing my best to get the woman's attention. It didn't take a genius to realize that she didn't want to acknowledge me, never mind open the door. After all, the shop wasn't due to open for some time yet, but I was beside myself with excitement and just couldn't wait. I started banging on the glass door, smiling broadly to appear friendly, all the while gesturing for her to open the shop. Finally, after some time, my persistence paid off. The woman gave in, came up to the front of the building, and opened the door.

Instead of being frustrated with me, she was kind and willing to help. She soon found a real bargain for me. I pulled out all the crumpled bills from my jean pockets and smoothed them out on the counter. When I finished counting out the money, I had enough for the ticket, with twenty pence to spare. I was on cloud nine, this was all too wonderful! I bought that ticket immediately and left for Switzerland that very day.

I was so thrilled to be leaving, so energized and excited about God's miraculous provision that I didn't phone ahead. When I arrived in Switzerland, there was no one there to pick me up, so I had to make a phone call with the single coin in my possession. "Oh Lord, please let there be a person there to answer. This is my only coin!" I held my breath as the phone rang.

One ring, then another, then...

"Hello, YWAM Switzerland." The sweet sound of a voice on the other end! God had answered my prayer. I was able to arrange for a ride to the YWAM School in Lausanne. I arrived without a penny in my pocket, but that day for me was sheer exhilaration. I felt like a millionaire. God, my provider, had brought me there.

Many times I have been asked by others how the money got in my Bible, I tell them I don't know. What I do know, is that God is a miracle worker who can make a way where there is no way. However my experience over the years would indicate that the Father loves to provide through other believers. If God used a believer in my case, he sure was an angel to me!

Learning to Be Faithful in Little Things

The school was a fantastic experience, as was our outreach in Paris. When it was over, it was time to head to England, and then onwards to the Philippines. Before I flew back to Asia, I made a trip to Scotland where I was invited to preach in a few churches. I didn't have money to get there either. However, there was a young medical doctor I had met who had learned about my need. He agreed to loan me the money to get a bus to Scotland, which cost a whopping five-pounds sterling. When I returned to London, I knew I needed to give the five-pounds back to the doctor, but I didn't have any money to give him.

I returned to the Philippines without paying back my debt. What could I do? I had nothing to give him. I rejoined my team and continued with them in our work. That particular month, however, my support money didn't come in on time. I was particularly aware of my need that month, as my teammates had been making jokes about my clothes. I was still wearing the same ones I brought with me two years earlier on that first Asian Circle trip! I didn't think that their clothes were much better, but I did note it might be wise to update my wardrobe.

Without my support money, there was no way I could buy new clothes or pay for any of my other needs, so I asked the Lord why the money hadn't come in on time. Son, the issue is that you owed that five-pounds in London, and you've not paid it back.

I felt so convicted! "It's true, Lord," was all I could say. "I didn't pay him back, because I didn't have money to give. If you give me the five-pounds, I'll give it to him immediately."

And would you believe it? The next morning, there were two letters waiting for me in the mail. The first contained my missing support money, which had been held up in the mail for over a month. The envelope looked like a stamp collectors delight, it had been forwarded from country to country all over the world! Such a thing had never happened to me before.

In the second letter was a five-pound English note. Until this day, it's the only time I've received foreign cash in the mail. I thanked the Lord for his impeccable timing and miraculous provision. Right away, I sent off the money to my doctor friend in England to pay my debt. After paying my debt, I had enough money to buy myself, and my YWAM companions, some new clothes!

Those important lessons have stayed with me my whole life. God is looking to train and encourage us so that we walk in the light at all times, taking care of details, and not overlooking them like I overlooked the five-pound payment in London. I didn't have the money I needed then, but neither did I go to God and ask for it. As much as God loves to do big things, He doesn't overlook the little things. Because of His generous and loving provision, I have been able to travel the world and work as a missionary, and by His grace and creative provision, I will continue to do His work for the rest of my days.

Chapter 12: Pioneering YWAM in the Philippines

After our Asian Circle Tour concluded, Mike and I found ourselves back in the Philippines asking what shape our work would take. What should we do? was our big question. There was no shortage of ministry opportunities in the country. I was offered a job to become a youth pastor at the first church we visited upon our return to Manila. Ministry opportunities abounded.

"I don't think it's smart for us to just jump into ministry now that we're back," Mike confided. It was another sweltering day in the city, and we had spent the morning looking for a place to stay.

"I agree. We really need to hear from God what He thinks. Let's set aside time to pray and seek direction from Him before we agree to do anything."

We made a conscious decision not to take speaking engagements or to get involved in ministries, things we could have easily done, but which would have distracted us from the business of hearing from God. Soon after we made this decision, a missionary couple from Canada, who were leaving for a short furlough, asked us if we would house sit their home in Makati City, Manila. We decided to take those weeks to listen for direction. It was a very spiritually rich time, putting God first by trusting Him with our future, expecting Him to clearly direct us. God was our priority, not the work.

We didn't know it at the time, but God used all those hours and days of prayer to prepare us for many years of ministry that were to come. Day after day, we created space for God. Mike and I spent most of it reading from the Bible, praying together, and seeking God quietly on our own. We implemented all of the principles we had learned from Joy Dawson when we first prepared to come to the Philippines on our Asian Circle Tour. These principles are still a part of YWAM's core values, internationally. We trusted that God would speak and clearly direct us.

Several weeks later as the Canadian couple were about to return, neither Mike nor I had received any specific direction. We were alarmed, so we really dug our heels into prayer. We became desperate to hear from God.

"The couple will be back in a couple of days, Kel. That means we'll have to move out!"

Mike didn't have to remind me that we needed to figure out our plan very soon.

"We've dedicated hour upon hour to prayer over these last number of weeks and yet we have no answers and no direction about what to do!" Mike confessed. "I gotta admit, Kel, I'm getting uncomfortable." It was difficult to balance our conviction that God would speak with clarity, yet our reality was that we still had no clear direction.

Someone once said, "It's the desperate cry of faith that God responds to," and we sure got desperate. As the final days of our stay approached, two very clear words came to my mind.

"I was praying early this morning," I told Mike, "and two thoughts came to mind. The first was 'Student Ministry' and the second was the word 'Baguio'. Have you ever heard the word Baguio before, Mike?" I enquired.

"It's a city up north," Mike responded, "but I don't know much else."

We looked it up on a map. Sure enough Baguio was a city in the northern mountains of Luzon. Neither Mike nor I had ever been north of Manila before. We eventually learned that this northern region is much cooler than other parts of the Philippines, and that many wealthy families from Manila went to Baguio during the hot summer months.

I folded the map, went to the window, and looked out at the quiet street. "These are the biggest clues we've had yet. Are you thinking what I'm thinking?"

Mike nodded his head. "Let's do it!" he said enthusiastically. "Student ministry in Baguio it will be!" We were in our final couple of days in the house, and once the couple returned to Manila, we packed our bags and headed north to Baguio.

Baguio City

Neither Mike nor I knew anyone up there, but we did know that the parents of our good friend and leader, Ross Tooley, had been living there. As it turned out, they were planning on leaving Baguio soon. We spoke with the owner of the house that they had been renting, and we were soon given a lease for the small home. It was located on the side of steep, rugged mountains to the north of Baguio, in a small mining village called Tuding. After settling into our new place, we went to visit our neighbor, the only other house near to us on the steep, winding road.

The Soriano family had built a large facility where they were running a Christian orphanage. We shared that we had come to Baguio for the express purpose of starting a student ministry. "We want to connect with the young people of the city," Mike explained. "And share the Good News with as many as we can."

"Don," the father called to his oldest son inside the house. "You must come and meet these men at once!" Turning back to us, the man explained, "My son is a recent graduate from university and was president of the school's Christian Union. There are six universities in the city, and Don can introduce you to many of the students."

We met with Don and started to talk about our plan to do ministry among students in the city. "What we really need is a property we can rent that would be suitable for a coffee house," I explained. "From the property we can do all sorts of ministry and meet with people our age."

Don scratched his head, admitting he had tried to find a similar location when he was president of the Christian union at his university. He had been unsuccessful. After some more thinking, he snapped his fingers and said, "Mrs. Brady! Yes, you should go meet with her."

"Why her?" we asked.

"Well," said Don, "she enjoys an excellent reputation in the city, she is very well thought of and educated. She's been involved in real estate and with City Hall for many years. I've also been told that she has recently become involved in the charismatic movement. Her husband recently passed away, leaving her widowed with five children. He was a well-known American lawyer, so she's very well connected. If anyone could help you guys, she'd be the one! She lives nearby, you can go see her tomorrow if you like."

Mrs. Brady and the Pink House

Up the incredibly steep and winding hill we went, eager to get into student ministry as soon as possible. We turned off the street and onto the property just like our neighbor described. We were impressed by the size and modern design of the home which indicated Mrs. Brady was quite well to do.

We knocked on the front door. We waited. We knocked again and waited some more. Just as we were about to leave, the door opened slightly and a woman, clearly a household maid, stuck out her head. We told her we were there to see Mrs. Brady. "Wait here," was all she said and then closed the door behind her. We stood outside the door for five long minutes, wondering if Mrs. Brady was lost somewhere in the house. Maybe she had forgotten about us.

Finally the door opened, and there stood an elegantly dressed woman. There was an air of sophistication about Mrs. Brady. She smiled warmly and seemed quite willing to engage us in conversation.

"Yes, hello," she said. From the moment she spoke, we could tell she was not just an ordinary person. She carried herself authoritatively, sort of precisely. "What can I do for you?"

Mike looked at me, and I explained who we were and that we wanted help, if she could give it, to find a place where we could reach out to college-aged students in the city. Mrs. Brady said nothing at first. She merely looked at us. Just as I was about to explain more clearly what I meant, she asked us, with a sense of urgency in her voice, "Do you know anything about the Holy Spirit?" I looked at Mike and smiled. We both nodded our heads affirmatively.

"Do you speak in tongues?"

We smiled again. "Yes."

Her eyes brightened. "Can you show me in the Bible where this happens?"

Again we said yes. "Oh!" She exclaimed, "you've been sent by God! Please come in."

When we entered her house, she recounted the events of her household over the last few months. Hungry to go deeper into the things of God, she had started to hold all-night prayer meetings in her home from sun down until sun up once every month. Her five children, two maids, the gardener, her chauffeurs, and some members of her extended family would meet there and pray all through the night.

On one night of prayer, Mrs. Brady was alone in her room offering supplications to God. "Suddenly," she said, "it was as though the sun started to shine like a laser beam down upon me. The room was filled with unnatural light in the middle of the night. The light came right through the roof of the house, into my bedroom where I was seeking the Lord. It shone like a flashlight, a beam of white light right onto my head."

She looked at us and blushed. "Then there was a tremendous power, like electricity. It repeatedly surged up and down through my body, from the top of my head down to my feet. I should have been terrified but I wasn't. I felt a peace like I had never experienced before, and when the light started to fade, I started to speak these funny words, without quite knowing what was going on. I had never heard those words before and I couldn't understand them."

Mike gave me a knowing look. "Later," she continued, "after that experience, I told other Christians about it, asking if anyone had similar experiences. A friend said that perhaps I had been baptized in the Holy Sprit."

She looked from Mike to myself, the look of a woman with a riddle too great for her to solve herself. "My daughter, Christine, told her friends at school about my experience and they were intrigued. They told Christine they wanted to come to the house to get a language they, too, could speak without having to study at school. I had no explanation, so I asked Christine to give me two weeks before she invited them home. I wanted to have time enough to understand what had taken place."

Mrs. Brady explained that it was nearing the end of that period of time. Almost two whole weeks had gone by, and she was no closer to solving the riddle. She had half a dozen students planning to come over the following night, and she had no explanation. She had locked herself away in her room, fasting and praying, waiting for God to intervene and tell her what she needed to know. She told her maids that she didn't want to see anyone or go anywhere. She would not accept any phone calls or interruptions until she had an answer from God.

Right at the end of her three days of fasting and prayer, here we were, on her doorstep. We had just come out of a similar time seeking God in prayer ourselves, waiting and asking God for help. Mike and I appreciated the desperate state Mrs. Brady was in. "Do you have a Bible?" we asked.

We immediately took her through a Bible study, showing her passages where the Holy Spirit had come upon the early church in the book of Acts. We showed her the passage of Pentecost when Peter stood before the people and boldly preached the gospel. We followed the footsteps of Paul across Europe and Asia Minor as we read about his many exciting encounters with people, and the surprising work of the Spirit. We read stories of how the early church was empowered by God, how they went out speaking in tongues, with holy boldness, preaching the gospel and seeing thousands come into the kingdom.

We circled back to the story of Peter and his encounter with the Roman Centurion. It told of how the Gentiles received the Spirit of God for themselves. We explained: "we've had experiences like this all over Asia. The Holy Spirit is available to us today, just like He was to the early church. He empowers today like He empowered Peter and Paul. That's what you experienced a few weeks ago."

Mrs. Brady was amazed. She had never read the passages before. She had only heard scriptures from her priest. (At that time, it wasn't common in the Catholic Church in the Philippines for parishioners to study the scriptures on their own). She was accustomed only to attending the service and having the priest tell her what she was supposed to know. After her prayer vigils and weeks of private time seeking God, Mrs. Brady was anxious to identify the truths in the Bible and have the proof from scripture that her experience was sane and true.

"I'd like you to be here tomorrow night when the students come and ask about the Holy Spirit." We assured her we'd be very happy to do that.

We held Bible studies in her home regularly after that. Young people from Catholic backgrounds came to the studies to get saved. Practicing English with visiting Westerners who knew about a secret language drew them in. Though we never did tell the students how to learn a language without any study, we did tell them how to have a personal relationship with the God who speaks to the hearts of men, the God who performs miracles.

The first night, six university students gathered. All six accepted Christ. When one of the students started to ask us a question, out of his mouth spilled the words of a unique language. The young man next to him looked at his friend strangely. He pointed and started to say something when the same thing happened to him. All six students were baptized in the Holy Spirit that night—that evening they all spoke in new tongues. Excited they asked us, "Can we please invite our friends tomorrow night?"

Matching their excitement measure for measure, Mrs. Brady responded, "My house is the Lord's house! You are most welcome to bring your friends." Then she turned to Mike and I and asked if we would be able to come. We told her we wouldn't miss it for the world! That next night, each of the six students brought a friend, and the same thing happened. They too accepted Christ, many being baptized in the Holy Spirit without any teaching about the Holy Spirit and without any teaching about speaking in tongues. We simply shared the gospel, and the Holy Spirit moved in power.

They too, left Mrs. Brady's home, excited and deeply changed by the love and power of God. Word started to spread throughout the student body of Baguio. Mrs. Brady was well known, well connected, and highly respected throughout the city. Students continued to come to her home, night after night, filling her house. Some nights the house was overcrowded—it got so full that people had to stand outside on upside-down boxes, in order to peer through the windows to hear what we had to say. God met many students there in a very personal way. The anointing of His Spirit was tangible and flowed generously in that home.

But God was not only interested in changing young students' lives. I remember sitting in the house during this period before a meeting and hearing Mrs. Brady talk with a friend on the phone. I could hear the voice of her friend talking excitedly. "Olga," she said, "I have heard reports that very strange things are occurring in your household. Do tell, what is happening?" The woman had a posh accent; she was the wife of a prominent judge.

Mrs. Brady just laughed and said, "You've got to come and find out about it yourself." That very night, the woman came; she accepted Christ, and received the Holy Spirit. Students, lawyers, judges, housewives, teachers, and other professionals were all amongst those who experienced this fresh move of God.

Mike and I continued to stay in Tuding, the little village just outside of Baguio, for many months. We rented a small house there, which sat quite high up on the side of a mountain. It was a fifteen-minute ride from this house into Baguio city where we were evangelizing.

Neither Mike nor I had a car, so we hopped aboard a Jeepney to make our daily trips into the city. After WWII ended, and the American military packed up and left, they left behind countless military Jeeps. Filipinos adopted these abandoned vehicles and started using them as sort of taxicabs, but with pre-designated routes. For a small charge, drivers picked up and dropped off people from place to place where demand was high, until a kind of bus system had begun to spring up all over the country. In order to distract from the memory of WWII, the Jeep owners painted over their vehicles in bright, crazy patterns.

People climbed inside and on top of these crazily painted, Jeep vehicles; they hung off the sides, and crammed in and on as many as could possibly fit. All the crazy colors, with the crazy number of people on board, made for a wonderfully amusing sight. This was the Filipino Jeepney. This mode of transport is still in use in the Philippines today. They have become a traditional mode of transportation for the public.

Since our transportation into the city had been sorted, we were able to run a "Jesus Festival" in Baguio's central park as a way to reach even more people for Christ. By the time the school year came to an end, Mike and I were nearly run off our feet between Bible studies at the different universities and the festival we held in Burnham Park. We literally ran from one Bible study to the next, taking turns leading and sharing from Scripture. We worked ceaselessly, teaching and establishing converts in their newfound faith.

During this dynamic and busy time of ministry, Mrs. Brady helped us to find a more permanent location than the little house we were renting on the side of the mountain. We found a home near Burnham Park, which we called the Pink House, due to its having been painted entirely pink. It was a great place to do all our discipleship and evangelism. Large and spacious, the Pink House was a lot less expensive and much more useful than the coffee house we had hoped to get could ever have been. Our whole team—which had grown quickly to around forty people—now had a superb place from which to operate. Thankfully, Mrs. Brady helped us to rent it on a long-term lease.

Just as our ministry was gaining a foothold, Mike, myself, and the team of newly converted young people we had working with us, began to encounter an increasing amount of supernatural opposition. For instance, a small hotel called the Ruff Inn near the Baguio airport was a place where a number of rumored supernatural healings were supposed to be happening. People flew in from all around the world to get healed by this particular Filipino 'faith healer' who happened to be using Ruff Inn. He reportedly diagnosed sicknesses and performed operations with his bare hands. It was said that he could remove growths without any pain or blood, leaving no scars. Many left believing they were healed, eventually he was exposed as a fraud. Someone took tissue he said he'd removed from a patient and had it sent to a lab for identification. It turned out to be tissue from a pig. As time went by we began to be more and more aware of spiritual opposition.

After a routine meeting in the Pink House, a young teenage girl, who had recently been converted, asked us to pray for her. As we prayed for her, she began to manifest a demonic spirit. She opened her mouth to speak, but out came a chilling hiss and terrible animalistic sounds. Knowing it was not the voice of the girl speaking, I demanded the spirit to identify itself.

The girl's mouth moved with contorted motions, and she spoke with a deep growling voice: "I'm the spirit of the black cat." She began writhing and making a scene—her face twisted into a sardonic expression. "I belong here, this is my home." Her voice took on a gravelly note, like she was speaking in a hideous purr.

"Not any more," I said. "I bind you in the name of Jesus Christ and command you to come out of her and go to the place of divine appointment." Instantly, the girl's voice grew soft, the distortion left her face, and she was wonderfully set free.

To gain some understanding as to how this spirit entered the girl in the first place, we asked her some questions about her past. I'd never heard of such a thing as a spirit of a black cat. "Do you know of a time when you might have encountered such a spirit before?" I began.

The girl nodded her head. "When I was young, I would wake during the night, and often I'd be freaked out by the sight of two big green eyes staring in at me from my bedroom windowsill. Because the cat was black, it's body blended into the night. All I could see were its green eyes staring at me. I'd lie there petrified. I was overcome by fear."

In her case, the fear became so great that it created an opening for a spirit to gain entrance into her life. But that night, by God's grace, she was completely delivered. The Holy Spirit overcame her fear and freed her from that darkness.

On another occasion, one of our short-term missionaries from Australia came under attack from another demonic spirit.

"Kel, Kel, where are you? You've gotta come quickly!" one of our team members was yelling. There was pandemonium in the house. People were yelling, and freaked out.

I came out of my room to find a major situation on my hands. This young man from Australia was lying on his back, flat on the floor with a crowd of other team members gathered around him.

"What on earth is going on?" I said as the team continued to pray and call out to God to save him. "It's Brian, it's Brian," one of the team yelled, he's just collapsed!

Brian was a strong young man in his early twenties. Without any warning, he had fallen to the floor, in one of the side rooms downstairs. His eyes had turned back, so that he was staring back at us with just the whites. A couple of the team members had searched for a pulse, but could not find one. It looked like he was not breathing.

We all prayed furiously, having no reason to believe that Brian was suffering from any actual medical condition, as he had not mentioned having any. Rather than call a hospital to wait for what would have been too long a time for help to come, we rebuked the spirit we were confident was attempting to claim Brian's life. One of our team members laid a Bible open on his chest, and with great urgency we all called on the God of the Bible to bring Brian back. We were becoming fervent and reaching desperation, when all of a sudden he made a slight movement. The whites of his eyes rolled and he began to focus. We watched breathless. Then, with a huge heave of his chest, his breathing resumed. Slowly he began to return to normal. A great round of applause, celebration, and thanksgiving to the Lord erupted from all of us missionaries!

A short while later, once he was sitting up and fully recovered from his ordeal, Brian told us that he had been out of his body. He said he was near the ceiling, watching us. He could hear us praying and calling out to God. He said, "when the Bible was opened and laid on the chest of my body, and everyone was praying, I could feel a power pulling me back to my body." That's when he stirred, and much to our huge relief, resumed his breathing and opened his eyes.

Spiritual warfare is as real as the ground we stand on. We were reminded again that day that our battle is not against flesh and blood, but against very real principalities and powers of darkness.

Brian was so thankful that we'd interceded on his behalf. "I'm glad to be back," he said with a smile on his face.

"Hallelujah, so are we brother." Hugs of joy, and many tears ensued.

In response to our prayers, and Mrs. Brady's deep longing for the Spirit, we established a long-term work among the students in Baguio. The Pink House was the place of many more stories of spiritual warfare and the work of God. It was the birthplace of Youth With A Mission in the Philippines.

Chapter 13: Divine Interruption

We continued to minister from Baguio City throughout 1972. The city served as the base of operations for all of our YWAM work, but I constantly had to travel into Manila to help our staff renew visas and do other types of paperwork.

In 1973, I was preparing to head back to New Zealand. Before I was due to leave, the pastor in Angeles City asked me to bring a team to share the Good News on the streets, and teach his people how to do evangelism.

One morning, as I sputtered down the road toward our ministry location in a beat-up old station wagon, I turned on the radio. To this day I'm not sure how it happened, but while I drove, the radio, which was playing music from a local station, all of a sudden was filled with static. Then just as quickly the static subsided and I found myself listening to the voice of a man who sounded like he was a pilot talking to ground control. From the way the conversation proceeded, some how I had mysteriously picked up a signal from the nearby Clark Air Force Base.

The radio crackled and I heard a voice over the airwaves speaking in an obvious American drawl, "Jones to command. We found something very interesting out here. We're going to take a closer look." Surprised to suddenly be listening to a military communiqué, I turned up the volume.

"Command to Jones, go ahead."

A few moments later I heard the pilot's voice again. "I'm looking at what appears to be a primitive jungle tribe on a high plateau, tucked behind a mountain. They're unlike anything I've ever seen."

The commander at the air base told the pilot to describe what he was seeing. "A-framed shelters," he said. "They look like teepees. I've never seen any thing like them over here."

My heart nearly stopped in my chest. I could hardly believe what I was I hearing? Had the military happened upon an unreached tribe, a native group having little or no contact with the outside world? As the pilot relayed coordinates over the radio, my brain started to buzz. The discovery of a previously unknown tribe would draw huge attention from sociologists, anthropologists, the military, the government, and the media. The thought of the tribe becoming specimens to be observed by scientists concerned me. They were people, human beings to be reached with the love of God!

I gripped the steering wheel and pulled a U-turn in the middle of the road. Something in my heart strongly resolved that the scientists in lab coats would not get to the tribe first. My ministry in Angeles City that day could wait. I knew I had to do something. I drove straight to the Air Force base.

I pulled my station wagon up to the guard post, and explained that I had come to see the commanding officer. I parked the old wagon in front of the main building. My Ford was spotted with rust and covered in dirt, it looked like a dinosaur parked next to the impressive military vehicles. It stood in stark contrast to their polished exteriors, all of them clean as whistles, with their wheels nearly reaching the height of the windows in my old Ford wagon. I couldn't help noticing that the military vehicles all had a beautiful thick tread on their tires too, unlike the bald tires on my old jalopy. I imagined that the jeep four-by-fours would be a delight to drive.

I stepped out of the station wagon, unsure yet what to say. I walked into the office asking God for wisdom under my breath. The commanding officer I was introduced to thrust out his arm. He grabbed my hand firmly and shook it. His uniform was sharp and spotless, with an impressive display of colorful pins decorating his chest. He was a serious man. Much like my old Ford station wagon, I looked rather out of place, silly even. My khaki pants were wrinkled, my Tee shirt was soaked in sweat, I had disheveled shoulder length hair, and was wearing flip-flops!

I cleared my throat and launched into an explanation of why I requested a meeting with him. I explained that I was a missionary, and that somehow my radio had intercepted a pilot's flyby description of a tribe up in the mountains. I requested permission to go and visit the tribe. To get to the mountains behind Clark Air Force base where the pilot had seen the tribe meant crossing into territory under the control of the US military. The officer's permission was essential if we were to make contact.

Thankfully I was meeting with a sympathetic man. He told me he had no problem if I went to visit the tribe. His response prompted a little more boldness. "Could you help us get to them?" I asked, thinking about the station wagon and how ill suited it was to get us even remotely close to a starting point from which we could enter the jungle.

The officer looked me over. "How could we do that?"

"If we were able to use one of your four wheel drive vehicles, we could travel across the rugged terrain and get as close to jungle as possible. It would save us days of trekking."

The officer was deep in thought for some time. He then looked out the window and spotted my station wagon. A smile threatened to invade the serious territory of his face. "I can lend you a jeep. One of my men will drive you as far as possible." After I thanked him, he asked if there was anything else we might need.

I was taken aback until a thought parachuted into my brain—an interjection of wisdom from God during my moment of dumbfoundedness. "Medical supplies," I spurted. "Do you have any spare first aid kits or medical supplies we could carry in with us?"

The officer thought this was a good idea, "As much as you can carry," he said. He sent me out of there that day with a generous amount of supplies: medicine, bandages, and ointments. We agreed on a day in the near future when I could come back and head into the mountains.

I left the Air Force base high on adrenaline, barreling down the road back toward Angeles City. What had just taken place? My whole day, my whole life it seemed, had suddenly changed. I woke up that morning with a clear plan for the day, prepared to help local churches minister to prostitutes and sailors, and just like that, God had rearranged my plans. In an instant, without warning or explanation, I was suddenly caught up in a new and surprising adventure. My plans had changed the moment I pulled that U-turn on the road.

I had to act fast. In a few days I would be heading out in a chauffeured military vehicle to trek deep into the jungle. The tribe, no doubt, would speak a specific mountain dialect, perhaps unlike any other, so I needed to find interpreters familiar with tribal dialects. I visited various churches to explain the unprecedented opportunity to meet a tribe no one had ever reached before, and to ask if there was anyone in their congregations who knew mountain languages.

I went from church to church, looking specifically for a couple of strong men who could share the weight of the supplies and keep pace with me. Not a single man volunteered to go with me into the mountains. Instead, two eighteen-year-old girls who were already pastoring a Pangasinan Assemblies of God church offered their help. They were totally excited about the tribe and explained they knew four different mountain languages between them.

I was blessed that they had offered, but I'd specifically been looking for guys. I had no idea what challenges or danger might lay ahead and didn't want to be slowed down by having girls on my team. Besides, I wasn't convinced they were up to the task. So I thanked the girls, and told them I would come back to them with a decision. Meanwhile, I continued to visit churches to round up some men who could accompany me. However, still no men were willing to go.

So I returned to the church where the young women had volunteered. They were over the moon with excitement that I had come back. I explained that there were a lot of unknown variables, that I wanted to make the trip as quickly as possible, and that we would venture into the jungle with only what supplies we could carry and our faith. The military had provided us with helpful instructions on to how to reach the tribe, and how long they estimated it would take us to find them by foot. The girls assured me they were very fast trekkers, and that they wouldn't slow me down. They were keen to go. I had a team!

Trekking Deep into the Jungle

A few days later, I headed out to the Air Force Base with Ivan Skinner (an Australian who had recently joined our YWAM team in the Philippines) and Angie and Carmelita, the two young pastors. A young officer helped us load up the jeep with our supplies. We each took a backpack complete with a sleeping mat, mosquito net, food, water, and all the medical supplies we could fit.

We needed much more water than we could carry with us into the hot, humid jungle. Because the area was so remote, we took a calculated risk that the jungle streams would be uncontaminated. The military officer drove us as far into the jungle as possible, then looked at us and said: "That's it guys, we're at the end of the trail, you're on your own from here."

Before we trekked off, the driver encouraged us that due to the small size of the tribe, we should be very careful not to miss them in the thick jungle or get ourselves lost.

In the back of my mind I was aware that it had been the habit of the military to drop men by parachute deep into the jungle to improve their survival skills. They were obviously better supplied and trained than we were for life in the jungle. Were we biting off more than we could chew? We didn't need reminding that we had no survival training at all and were completely trusting in the Lord to be our guide and protection.

The military knew about one primitive tribe that had a village on the way to the more isolated tribe that we hoped to meet. "Be warned," the officer had informed us, "a warrior from that tribe recently shot an out-station Air Force serviceman with bow and arrow!" We knew if we encountered this tribe on our trek, we should be thoroughly on our guard and watchful but at the same time happily assured that we were headed in the right direction.

We thanked the driver for the ride and for his obvious concern for us and stepped into the jungle.

We walked closely together. Small animal tracks were worn into the rain forest's floor, so we followed the little pathways they made in the general direction of the tribe. It would be impossible to make any real progress by cutting our own path through the jungle. The vegetation, thick and lush, burst from the red earth and spread dark green leaves like a canopy above us, but never so thick that we couldn't catch glimpses of the blue sky. Birds fluttered and squawked from high branches.

Every so often a plane would rip through the sky high above the trees that towered over us. "An F-5E Tiger!" Ivan shouted as our small expedition looked up through the trees. "I know the sound anywhere!"

"Doing recon, no doubt, like the pilot who spotted the tribe we're searching for," I said. "Let's keep walking."

The further we walked, the thicker the jungle grew around us. And then just as it seemed to get to its thickest, we would come into a clearing. In these areas there were patches of long grass we had to trek through. It was terrible stuff! Like long elephant grass in Africa, this grass was reed-like, coarse, and difficult to navigate through. The blades had razor sharp edges which easily cut through the skin. After the grass, the jungle would become dense with trees again.

We had come to a point in the trail that edged around the side of a mountain. Ahead of us was a steep incline, and across from the rock face, vibrant jungle surrounded us. The birds weren't the only creatures watching us. We had been trudging through the jungle for a few hours, quietly talking with each other as we went, and listening to the sounds of the jungle, when I stopped dead in my tracks. Ivan bumped into me. He asked what was the matter; I held my finger to my mouth.

We were at the base of a sharp incline, at the top of the hill stood a warrior. He was naked except for a covering of vines wrapped around his loins. Across his chest was a strap holding a furrow of arrows to his back, they stretched a full foot taller than his head. In his right hand he held a bow two feet taller than himself.

"I think we'd better stop," whispered Angie.

I agreed. The four of us stood closely together in a group. We looked up at the warrior, and he looked down at us. A long moment passed, quiet except for the buzz of insects in the jungle and the blood pounding in our ears.

"You need to say something," Angie whispered.

I turned to her and held out my hands. "You know the mountain languages, not me. I think you should say something."

Angie frowned and nodded toward the man. "It's not us he's interested in, Kel, it's you."

I looked up the hill and saw that she was right. The warrior's gaze was fixed directly on me. I didn't have a clue what to say, so after a moment of dead silence and a desperate inward cry for help, a thought occurred to me.

"Give me a demonstration with your bow," I yelled out. Angie was quick to translate it in a language that, fortunately for us, he understood. Immediately a smile arched across the warrior's face. He lifted a machete—which I had not yet seen—from a sheath strapped to his leg. He held it in the air. Ivan looked at me nervously. We sighed with relief as he turned to the side of the trail and lopped off the stalk of a small banana palm.

With the end of the machete, he dug into the dirt and planted the banana stalk in the middle of the trail. He stomped the red earth until it was packed around the stalk so that it stood straight out of the ground. The warrior then ran down the steep mountain side to where we stood. From our viewpoint, the distance made the banana stalk look like the size of a toothpick. It was immediately evident to us that the warrior had set a target that was impossible for him to hit.

"Impossible," I said under my breath. "Let's move closer." I didn't want the warrior to miss his target and be embarrassed or feel any shame for missing in front of complete strangers. The warrior refused to move even one step closer. I had challenged him from the spot where we first saw each other, so that was the very spot from where he would shoot.

We stepped back to give him some room. He was short compared to Ivan and I, at least by a full foot. We were both around six feet tall, he barely scratched five. The bow he held in his hand reached from the ground up to the top of my head. The man pulled one of the tall arrows from his quiver, lifted the bow and took aim.

Before any of us could think, the warrior had unleashed the arrow. The whoosh of the arrow was followed by a sharp crack, the sound of an arrow hitting rock. We looked up the mountain trail and saw the arrow had pierced the side of the mountain.

The banana stalk had not moved. I looked at the warrior and smiled. "You missed." Angie was being careful to translate everything I said.

He just shook his head, "No."

"Look at the palm," I said, pointing up the hill. "It's still standing."

He shook his head again, smiled, and beckoned me to follow. So up the trail we went. When we got to the banana stalk, Ivan and I were stunned. It stood upright but at the top of the stalk was a perfectly round hole, the width of the arrowhead. The arrow had pierced the stalk so fast we never saw the banana stalk move! The warrior was an incredibly accurate marksman.

I told him how impressed I was and made it clear I thought he was an excellent shot. Night was fast approaching and our team needed a place to stay. "I would feel very safe if I could stay in your hut tonight," I said. A broad grin of agreement crept over the man's face.

The man was happy to offer hospitality. He told us how proud he would be to host foreigners in his hut. The tribe would be impressed, and his status would increase. He led us further through the jungle to his village. Ivan and I stayed in his hut for the night, and Angie and Carmelita slept in another hut in the village. We slept soundly, thankful the warrior shot his arrow at the banana target rather than at us. We were happy to have a roof over our heads, to be dry and to have experienced wonderful hospitality from the tribe.

Chapter 14: The Tribe on the Plateau

We woke before sunrise the next morning, leaving the tribe in the dark. We wanted to get as much distance on our trek toward the unreached tribe as we could before the sun was up. We walked in the cool dark of the early morning, with thoughts of the mountain tribe we were looking for powering our every step.

We weren't too far from our hosts of the previous night when we met another warrior who had been out hunting through the night. He had been successful. He proudly brought us to his prize: a twenty-foot long python. This monster of a snake had a body as thick around as my thigh. The warrior was starting a fire to cook some of the snake meat, which he offered to us. We respectfully declined, thankful that our stomachs were full on rice that the tribe had fed us before we set off on our trek for the day.

We watched as he hacked through flesh, sinew, and cartilage. The warrior cut the python in segments so he could carry the meat back to his tribe. The snake was so heavy that he was only able to shoulder a few segments of the massive reptile at a time. He would have to return with other members of the tribe for the rest. We parted ways and moved on with our mission, leaving the warrior as he carried chunks of snake back to his tribe.

We continued our trek through the jungle, working our way up the side of a mountain, swatting at mosquitoes and suffering in the heat as the sun continued to arc its way higher in the sky. The trees offered shade, but the tropical heat and humidity took their toll. We slowed in the heat of the afternoon, humming songs under our breath in concert with the symphony of insects, birds, and other invisible creatures hidden in the trees.

When we reached the top of this mountain, we dropped our packs and worked out the kinks in our backs. We were all tired and sore. I was particularly proud of the girls who'd definitely lived up to their word: they were excellent trekkers. I was soaked to the bone with sweat. The lot of us were relieved to have made it to the highest summit where, according to the sketchy outline we had from the military and the coordinates of the pilot, we should be able to see the lost tribe.

We looked in every direction from our vantage point at the top of the mountain. We couldn't see a thing. There was no indication of a tribe, and the plateau was nowhere in sight.

Could it be that we had made some wrong turn in the morning dark? Were all those hours of trudging through the jungle toward the mountain's summit spent walking in the wrong direction? The officer had warned us to be very careful, that it was easy to be turned around in the thick vegetation. We walked around to the other side of the mountain, to the back of its slope, to see what we could see.

To our great relief, when we ventured to the other side of the peak, there it was: a long plateau, almost three football fields in length! It was flat and barren in contrast with the steep mountainside, which was covered in thick vegetation. The plateau looked almost man-made, as if it had been cleared by a bulldozer and flattened by heavy machinery. The site was perfect for a village. We caught our breath, as we surveyed the scene.

"Smoke!" Carmelita exclaimed. She was right. Wisps of smoke were clear against the backdrop of blue sky. Just as the pilot described, we saw A-framed huts that looked like Native American teepees, standing a few feet apart, all over the plateau. We looked at each other in excitement, and thanked God that we had found our way to the tribe.

"Listen," said Ivan. "Can you hear that?"

We all went silent and strained to listen. "Voices!" But all we could hear was the distant drone of people talking, nothing was distinguishable.

"We'll have to get closer."

We decided that it wasn't wise for Ivan and I to trek down the valley and up to the plateau on the other peak. If the tribe had little or no contact with outsiders, they might be terrified at the sight of six-foot tall visiting ghosts. Or, worse yet, become militant and attack us.

After praying together about what to do we agreed to send the girls alone to make contact with the tribe. They were to try out the mountain dialects they knew, and hope for the best. Since we had no knowledge of the tribe, whether they would be open or closed to outsiders, we wanted to limit their perception of hostility as much as possible.

Ivan and I stood at our lookout point, and sent off the two young Filipino pastors to make first contact by themselves. I told the girls if they were able to speak their language, that they should try to find out whatever information about the tribe they could. Did the tribe have witch doctors? Were they open to have some visitors bring them some Good News?

We worked out a simple system of communication before the girls left, a way to relay to us if we should follow after them. If the tribe was open to foreigners and wanted us to visit, the girls would find a spot where we could clearly see them on the plateau and wave their arms above their heads. If they didn't wave their arms for us to follow, if they left the plateau or had to run, then we would know that the tribe was hostile, and that the two brave boys would need to pray hard! It was essential that our communication method be uncomplicated. While we knew we were running a definite risk, each of us had decided ahead of time that we were willing to pay whatever price God asked of us.

Ivan and I sat down to intercede for the safety of our translators and for their speedy access to the tribe. It was about an hour and a half's journey for Angie and Carmelita to get down the mountain, across a river, and up to the side of the adjacent mountain where the tribe dwelt on the plateau. We watched until they disappeared over the slope below us, into the jungle. We were concerned about them having to cross the river at the bottom by themselves, but they told us not to worry, they were totally confident God would help them find a way across. Before leaving us, they said, "Don't worry about us you guys, we're used to this, we're mountain hikers!"

"They sure got that right," I said to Ivan. "They've had no problem keeping up with us, or carrying their packs." I was glad God added them to our team, and broke my mental notion about only looking for male interpreters. I said to Ivan, "They're probably better than any guys who might have joined our team, don't you think?" Ivan was quick to nod his agreement.

These two young woman pastors were simply the best. They were incredibly adventurous young women of God. I had become increasingly impressed with their athleticism and energy levels. Never once did they ask us to slow down or suggest we take a break. Neither complained but always, they had this cheerful, positive outlook.

We prayed fervently for their safety and success as we sat there on the mountaintop. Every now and then we'd be rewarded with glimpses of the pair as they navigated their way up the steep mountainside opposite. They meandered their way through thick vegetation and over endless rocky outcrops. Little by little, they gained in elevation. Our anticipation increased as the gap closed between them and the tribe.

They reached the edge of the plateau. We could hear their voices, shouting to the tribe. The entire tribe reacted by running away from our girls. Not an encouraging sight. The girls took a few steps closer but when they did, the whole tribe took a few steps back. Our two women were just over five feet tall, yet they were still taller than the tallest men of the tribe, they were nearly a foot taller than the women. The tribe watched them closely, making sure they kept their distance.

After a few attempts, the girls stopped trying to approach the tribe. Instead, they called out from where they were. Systematically, they tested each of the dialects they knew for recognition. After a few different pronouncements in each different tongue; eventually, one of the men in the tribe shouted back. The women were able to communicate! I looked at Ivan and smiled. We couldn't understand what was said, but we knew the longer they spoke, the better our chances of visiting the tribe.

We watched as the girls shouted to the tribe and the tribe shouted back. The distance between the two groups was about the width of a football field. The tribe edged a little closer to the women. Shivers went up my spine as the girls turned and raised their arms, waving energetically. The tribe was open and willing to meet with us.

Chapter 15: The Wuh Tribe Continued...

Ivan and I hit the trail. What an opportunity! We didn't speak much as we made our way down the steep rocky mountainside through the clinging vines and endless ruts. We made our way across the river and up the slope toward the plateau. I could see by the bounce in Ivan's step; he was as excited as I. The fatigue of the many hours of trekking over rivers and up and down mountains seemed to fall from us the closer we got to the tribe. Adrenalin was feeding our excitement with each step, then all of a sudden we emerged onto the plateau.

When the tribe saw us they gasped. We towered over each man and woman, nearly two feet taller than most of them. In terms of size, they could easily have been mistaken for a tribe of people with dwarfism, except that their small bodies were perfectly proportional. The children, brown eyes wide in astonishment, screamed and ran behind their parents' backs. Afraid but curious, they poked out their heads from behind their parents' legs to observe the strange white visitors.

Ivan and I matched their astonishment measure for measure. I had never seen a group of people like this tribe. Each member of the tribe seemed completely bald. Men, women, and children alike—all with shaved scalps. We estimated about sixty people in total.

They stood in silence watching us. All were very short, and mostly naked, except a few warriors who wore vine loincloths. The women were decorated with braided vines, which adorned their necks, ankles and wrists.

There were about twenty huts on the plateau. Each of these was around eight feet tall, made out of long tree branches, and woven together with dried grasses. Inside the huts we could see mounds of earth piled in rows that were covered in leaves and soft grasses. These were the natural mattresses upon which they slept.

We greeted the girls and asked them what information they had gathered about the tribe. The tribe watched our every move. In similar fashion to the girls' arrival, the tribe stood apart from us by about thirty yards. They were far enough away to be safe, but close enough to satisfy their curiosity. I wanted to get closer to speak with them, so I slowly attempted to take steps toward them. When I did, they started slowly moving backward, making it clear they didn't want me to come any closer. They had established the distance from which I could communicate to them. This meant I would definitely have to shout. They were interested in us, but also clearly afraid of the giants with white skin standing in front of them.

"Do they have witch doctors?" I asked. I knew that if the tribe were under the influence of witch doctors, they'd likely have a great fear of the supernatural realm.

"Yes," said the girls. "We already asked. They have two."

We stood there, all of us, just looking at each other. Mindful, silent. Finally, Angie turned to me and said, "Kel, you need to say something."

I stood there perplexed. I looked around their plateau and realized just how foreign I was to the tribe, and how foreign they were to me. To them, we must have looked like we were from outer space! None of us had ever encountered people this primitive. There was no sign of anything from the outside world. They wore practically no covering or footwear. They had no technology of any kind, or any of the other conveniences many of us take for granted in our modern world. There was nothing but small huts, mounds of earth, fire pits and jungle.

What was I going to say to this wide-eyed tribe of totally freaked out people? I found myself totally lost for words. I didn't have a clue what to say or do.

My excitement to reach this tribe was suddenly transformed into an overwhelming sense of desperation. I did the only thing I knew to do, I screamed out to God in the silence of my heart a single word: Help!

The moment that I cried out to God, a passage of scripture was dropped into the secret chamber of my heart: Romans One, Kel. As those words echoed through my mind, I scrambled to remember the passage I had studied so many times before. And as my heart pounded in the growing silence while the tribe and my small team waited for me to speak, I was reminded of Paul's words to the early church in Rome:

Forever since the world was created, people have seen the earth and sky. Through everything God made, they can clearly see His invisible qualities—His eternal power and divine nature. So they have no excuse for not knowing God. Romans 1:20 (New Living Translation).

God has given a revelation of Himself to every man, woman, and child through the creation that is all around us, a creation that always has, and always will, reveal His greatness, creativity, and power.

I had a sermon!

Chapter 16: The Good God

I looked over the tribe. Men, women and small children were all staring at me intently, expecting me to say something. I had an attentive audience. They stood on the fringes of the plateau, still maintaining a safe distance.

The words from the first chapter of Romans resonated in my heart, and pumped like blood through my body. How do I make those words plain to these people? This tribe has never even seen paper or a book before. If they've never seen paper, how in the world can they relate to the Bible?

I looked at my surroundings. Beyond the plateau I could see the outline of trees and banana palms moving slightly in the breeze. The fruit trees inspired me with how to implement this sermon fresh from Heaven. Still yelling to the group from across the distance, I pointed at the banana palms and loudly said, "Sugging!" the tribe's word for banana. The tribe looked over at the leafy palm with a stock of bananas. Through one of the translators I yelled, "Do you like Sugging?"

The tribe looked at each other, their brows furrowed in thought. They looked back at me, and in unison said "Wuh!" nodding their heads Yes.

"Does Sugging taste good?" "Wuh," they replied again nodding their heads in agreement.

"Do you know who made Sugging?" I asked.

The tribe looked at each other and then at me, and shook their heads from side to side. "Wuh!" They replied. They did not.

"God did!" I announced. "Isn't God good?"

"Wuh!" the tribe exclaimed in agreement.

I looked around me. What other things from creation, from their daily lives, could the tribe relate to? I noticed a clay pot filled with water, sitting over a fire. I stepped toward it and my stomach churned. A dark mysterious soup boiled over the flames and I looked more closely inside the cauldron. Stubbed snouts and mouse like ears pointed out of the water. Singed bat fur that had separated from the leathery wings floated in the black inky water. The creatures may have looked like little devils to me, but I understood that these nocturnal animals were one of this tribe's main sources of protein.

I swallowed the lump in my throat and turned to the tribe. "Do you like bats?" I yelled.

"Wuh!" the tribe exclaimed.

"Do bats taste good?"

"Wuh," they all agreed again.

"Do you know who made bats?" I asked.

The tribe shook their heads. They did not know.

"God did!" I said loudly, pointing to the sky. "Isn't God good?"

"Wuh," the tribe agreed.

I looked over my shoulder and pointed at the face of the mountain behind me, from where earlier Ivan, myself, and the two girls had come. "Do you like the mountain?" I asked.

"Wuh" the tribe responded that they did.

"And do you know who made the mountain?"

"Wuh," replied the tribe, this time smiling.

"God did," I smiled back. And with a chuckle said, "Isn't God good?"

The whole tribe agreed.

We stood on that mountain for some time. I continued to point out different parts of the created world around us, the beautiful inventory of things that God had made: mountain rice, delicious snake meat, berries, and the birds the tribe hunted for food. My message sounded more like a field trip for a botany class than a sermon, but we got into a wonderful rhythm of observing and agreeing with each other that God had made a beautiful world for the tribe to live in, with its fruit, rice, birds, plants, sky, land, and geography. "Isn't God good?" I would say, asking the question they had started to expect me to ask, and they would say "Wuh!"

When I ran out of fruits, vegetables, and other natural illustrations, I frowned and looked over the tribe. God's revealed character through nature was only part of Paul's emphasis in the chapter as I thought further about the passage:

For the wrath of God is revealed from Heaven against all ungodliness and wickedness of those who by their wickedness suppress the truth... for though they knew God, they did not honor him as God or give thanks to him, but they became futile in their thinking, and their senseless minds... exchanged the glory of the immortal God for images resembling a mortal human being or birds or four-footed animals or reptiles. Romans 1:18-23 (New Revised Standard Version)

As I watched the tribe, surveying its men, women, the little children that gathered around their parents, after a long moment I bowed my head. "One day, you made the Good God sad," I said. When they heard the translation, the whole tribe became perplexed. They frowned too. "Do you know what you did to make the Good God sad?" I asked.

"Wuh" they replied, shaking their heads from side to side, they did not know.

I shared with them the passage from Romans chapter one, where Paul lists sins common to all people everywhere. Since the Bible says that sin is common to all men, I suspected the tribe would identify with some of the sins I mentioned: covetousness, malice, murder, strife, craftiness, gossip, drunkenness, foolishness, fornication, adultery, and on I went.

The sun was still high in the sky as the girls slowly translated what I said, but darkness began to pass over the faces of the tribe members. We could tell, even though we were perhaps 90 feet or so away, that the happy faces from moments earlier were now suddenly sad. They were still hesitant to come near us, but we could see as they turned to consult each other with what we had shared that the tribe was becoming emotional.

One of the men in the tribe, who we later learned was the chief, had really taken the words we'd just shared to heart. From the look on his troubled face, he was agreeing with the fact that his tribe was indeed guilty of such things. We could sense a great internal struggle was going on in the man and before long his quivering chin gave way to a terrible loud cry. He was deeply convicted and began to unburden himself with loud wailing that came from deep within. He was broken over his sin.

A man beside him started to cry, he too was coming under the same conviction. Then another. And another. A great wave of conviction enveloped the entire tribe. Their children, perhaps out of the same conviction, perhaps merely frightened by the fact that their parents were openly weeping, began to cry too. Very soon the whole tribe was openly crying. Raw emotion rippled through the whole group until there wasn't a single dry eye.

I looked at Ivan and the girls. Their faces said it all. Their eyes were also brimming with tears. We had shared a simple message with them about the goodness of God and our inherent failure as people to live up to his standard of goodness. The whole tribe fell under the conviction of the Holy Spirit. God had truly prepared the hearts of the tribe to receive the Good News that day.

The weeping among the tribe started to reach a fever pitch. The conviction seemed to overwhelm them and as a group they started to really wail. Before we lost them into a fit of despair, I held up my hands to tell them the really good news. The girls had to yell loudly above the fray to get the tribe's attention to ask if I could tell them what God had done for them.

I told the tribe the story of the cross, how God had come down in the person of a man and taken upon Himself the guilt and sin of all people. I told them that the guilt of their sin was nailed to a tree through the crucifixion of Christ so that all sin could be forgiven and all people could be restored into relationship with the Good God.

When the tribe heard this news, they wailed even louder. They now understood their guilt, so they asked us, "Why would a Good God want to forgive us?"

We interrupted their wailing for the second time, and asked them if they wanted God to forgive them of all their sins, if they wanted to invite the Good God into their hearts so they could be right with Him. Once again it took a considerable effort for the girls to quiet the emotional tribe.

"The Good God wants to be your God. He wants to forgive you," I assured the tribe. "He wants you to be His children." With my heart pounding in my chest I asked the tribe, "Do you want to be God's children?"

"Wuh!" the tribe said as if in one voice. They wanted to be God's children very much. We led the whole tribe into a prayer of repentance and salvation. Everyone in the tribe accepted Christ, except for one of the two witch doctors.

Chapter 17: Confronting the Lies of the Witch Doctors

The moment they said yes to Christ, their sins were lifted and they were now free to be God's kids. It was like Heaven was opened above them, and a profound excitement and jubilation settled over them. Instantly their fear was gone, their faces wet with tears, now revealed a look of astonished joy. My spirit soared with them, just as it had all those years ago when, on my knees at the Memorial Hall in New Zealand, I had first truly professed my faith in Christ Jesus.

One by one, members of the tribe began to break rank and walk in a V-shape toward us. They approached us, looked up into our faces, then at each other, talking amongst themselves, laughing in amazement, pointing out our strange features.

When they came near, their body odor overwhelmed us. The stench of dirt, sweat, urine and feces was so putrid it almost knocked us off our feet.

I was thankful that their fear had kept them away while I was preaching the Good News. Had I been standing amongst them, it would have been a major challenge for me to overcome my repulsion from that stench. "Thank you Jesus, thank you," I whispered under my breath. All the preaching from a distance had been the grace of God!

Now that they'd committed their lives to the Good God, their fear of us was gone. They gathered all around us. I was able to look at them closely. I realized that they weren't bald at all; they had something like pottery on their heads. What once looked like dark-skinned, hairless heads materialized into clay-like helmets. As the girls started to talk to the tribe, they learned that the witch doctors had told the people there were evil spirits in the water, so the tribe refused to wash or bathe in the nearby river. They were ruled by the fear that if they went in the water, the spirits would kill them.

It was unclear if the witch doctors had only recently come up with this fearful tale, but it was apparent that no one in the tribe had bathed for a number of months. No wonder they smelled so terrible! The clay caps, then, were on their heads not because of some tribal ritual or as cultural decoration, it was an accumulation of dirt and earth after months of sleeping on the ground and not bathing. As it turned out, their short frizzy hair had mixed with clay and sweat to form a crust on their heads, hardened by the punishing tropical sun.

One of the men who stood near to me was scratching at the hardened clay helmet on his head. I went up to him and knocked on his scalp. It was like knocking on a terra cotta pot. The hollow echo reverberated off his head. The man was deeply distressed and clawed at his head, but all his efforts to gain any relief appeared fruitless. I attempted to break off a piece of the hardened clay near his neckline where the clay looked more fragile and thin. After successfully breaking off a small piece, I was horrified to discover white maggots wriggling between his scalp and the dirt helmet. My stomach cringed at the sight of bugs crawling on his scalp. No wonder his head was so itchy! We proceeded to do what we could to help him.

I asked the tribe, "Do you drink water?"

"Wuh," they replied.

"Where do you get it from?" I wondered.

"From the river," they said.

"Well, wouldn't you be swallowing the evil spirits if they were really in the water?"

After thinking about this for a while they said they'd never thought about that.

"You don't need to fear the water," I assured them. "The Good God made water good too. There are no evil spirits in the water. And it tastes pretty good, doesn't it?"

"Wuh."

The tribe became angry that they had been tricked and so they started to taunt the one witch doctor who had refused to accept Christ. It became apparent that the witch doctors would offer blood sacrifices to the spirits, go into trances and become filled with demons. They'd often make up stories and lies to force the people to live in superstition, fear and terror. It wasn't long before the tribe excommunicated the unbelieving witch doctor and sent him away.

The overall health of the people was poor. Many in the tribe had ulcers and infected open wounds, filled with puss and blood. It was horrible to look at. I couldn't imagine how much pain they were in. Though the tribe had beautiful brown eyes, many squinted at us painfully, blinking through swollen lids with pink, infected eyes. Flies buzzed around the clan, festering in their stench. The little kids, tired of swatting them away, itched their dry and irritated skin. So many of their health conditions could be easily treated with a basic regime of daily hygiene.

I then noticed their feet. The ankle area of the foot looked normal, but quickly the rest of the foot became unusually wide and wedge shaped. Because their feet grew so wide, they had enormous spaces between their toes. No normal shoe or sandal would ever fit them! Regularly, the people would traverse the treacherous and slippery mountain trails and rock faces often carrying heavy loads on their backs. I suspected that, as the people climbed up and down this steep region in their bare feet, their toes would need to grip into the mountainside and the constant strain of climbing would spread their toes apart. The tribe looked at our shoes, pointed and gawked. I slipped off my shoes and socks and they gasped when they saw my white, and by comparison, narrow, funny-looking feet.

Children approached us, cautiously at first, and touched at the hair on our arms, something they'd never seen before. They would pull on it to see any of it would come off! They rubbed our skin and asked Angie and Carmelita, "Why won't the white paint on the men rub off?" The whole tribe examined us and seemed completely baffled. Never having seen white men before, they were convinced we'd painted our skin white.

Neither had they ever encountered fabric before, or seen hair that wasn't black like theirs. Then they discovered our eyes were blue, this was altogether intriguing. One by one they would come right inside our space until they were only inches away just to peer into our eyes. They were simply astonished. The whole tribe jostled for position, they all wanted to be close enough to touch our clothes, hair, skin and look into our eyes.

The tribe asked us to teach them as much about the Good God as we could. We told them we were happy to teach them all that we knew, but first things first. We wanted to clean them up and relieve them of their physical discomfort. The four of us opened up our bags of medical supplies and went to work.

The first man I attended to had a horrible ulcer on his arm. It was swollen and ugly, filled with dirt, and puss slowly dripped off his elbow. I dipped some hydrogen peroxide on a cotton swab and started to clean out the wound. He winced as the cool liquid touched his skin; through one of the girls I assured him it would not be painful. When I was finished cleaning and treating the wound I covered it with a three inch wide Band-Aid supplied by the military. When other members of the tribe saw the funny, brown adhesive sticking to his skin, they reached at it, ran their fingers along it, and started digging through our bag of medical supplies.

Before I knew what they were up to, they had scooped up all the Band-Aids they could and started to fix them all over their skin. One member of the tribe had one, so they all wanted one. This is how life worked up on the plateau: all for one and one for all. In no time, all of our Band-Aids were used up, stuck to foreheads, calf muscles, shoulders, stomachs, buttocks and feet. Not a single one covered the cleaned wounds. Instead, they were strange and new decorations for the tribe. It was clear that we would soon have to make a trip back to the military base to get more supplies. More importantly, we needed to communicate to the tribe the purpose of Band-Aids.

We wanted to help the tribe get out of the nasty, uncomfortable potted helmets. After a few days, Ivan and I made a quick trip to the military base. We gave a brief report to the military authorities, and requested more medical supplies. We got hold of some empty four-gallon cans, and turned them into buckets, filled them with all the supplies we could and hauled them back out to the plateau. On our arrival, we emptied our supplies and proceeded to fill the metal buckets with water, which we then heated over a fire. The two girls pulled out bars of soap and some shampoo and with the aid of the hot water, began to tend to their imprisoned scalps. They wet the tribe members' heads with water and slowly lathered the soap into their scalps. The project was a lot more difficult than we'd imagined. The girls taught them how to wash without getting soap in their eyes, lest the tribe think the stinging pain of the soap in their eye sockets was the work of some dark or evil spirit.

At first, they even tried to eat the bars of soap, thinking they were food! Some days later, when the whole tribe was clean and washed, they truly looked like the new creations they had become: clean on the inside and out! The tribe laughed and touched each other's soft, clean hair, visible for the first time in a long while.

The tribe welcomed us like we were long lost family. They loved us and we loved them. Ivan and I stayed with the tribe for just over a week. I had a mission base to run, though, so Ivan and I headed out. The two girls remained, and became the tribe's pastors.

As we walked back through the jungle, back toward our regular ministry and to the flurry of civilization, Ivan and I reflected on our time with the tribe, amazed at what God had done.

"They are so completely different from us, with such different experiences," I said as we weaved through trees on our way back toward the military base. "It really is true isn't it Ivan, that by His grace God dramatically draws all people to Himself. The Good God I introduced the tribe to is the same God who is the Good Shepherd who lays down His life for His sheep."

Ivan stopped and pulled out the canteen from his backpack. We both took a sip of water and caught our breath. "It's so easy to take all that we know for granted, isn't it?" he mused. "I mean, about God, but also the basic meaning of life. It was crazy to see how the witch doctors deceived the tribe and how that deception affected their health. But God loves them so much that He cares for both their spiritual and their physical needs."

As we continued to walk through the jungle, we talked about all the ways we could see how God provided for the needs of the Wuh tribe.

"He will feed His flock like a shepherd," I said as the sun started to drop below the height of the trees. "He will carry the lambs in his arms, holding them close to His heart. He will gently lead the mother sheep with their young."

As the forest around us grew darker and cooler, our hearts were warmed with thankfulness to God. Little did we know that as we left the tribe behind us, it would be the last time Ivan or I would ever see them on this side of glory.

Chapter 18: Mount Pinatubo

The tribe was illiterate and had never recorded their language. The girls taught them Bible stories which the tribe quickly memorized. Their memory capacity was simply remarkable. The girls sent me updates on the tribe's progress through occasional letters to Baguio. There were no postal services in the jungle, so letters could only be sent when the girls come out for supplies.

The tribe memorized things through story, and after repeating a verse a few times to the group, that verse would be engrained in their memory. The whole tribe began to learn large segments of scripture by heart while the two girls established a program to teach the tribe to read and write. They had no pens, paper or chalkboards from which to teach, so sticks were used to form the letters of the alphabet in the ground.

The village play area became their classroom. The whole tribe would gather together to learn, old and young alike. We were able to contact The Philippines Bible Society and discovered they already had some small portions of scripture in the language the tribe spoke. So these portions of the Bible soon became their textbooks.

My connection with the tribe, however, would not last. The New People's Army (NPA), a militia wing of the Communist Party in the Philippines, swept across the country hiding in mountainous jungle regions where it planned out attacks on police and government. They were an extreme group of terrorists who wanted to take political control of the country.

In the early years of the 1970's, the political climate in the Philippines was volatile and would remain so all through the presidency of Ferdinand Marcos, who ruled the country with a heavy hand until 1986. Marcos declared martial law, silenced the press, and squashed all opposition to his power. The Communist NPA actively opposed him until, eventually, the people rose up against his authority. Marcos fled the country and was exiled to Hawaii.

Our two young pastors, who had discipled the tribe, were forced to leave the area because of the danger the NPA presented. Both bandits and the NPA would hide out in the mountainous area around Clark Air Force base, successfully hiding from government forces and launching attacks and raids. The United States military finally decided it had become too risky for the two pastors to continue to go in and out. They could easily become targets and held for ransom; or worse yet, be raped and killed. Permission to enter the tribe through the military controlled land was revoked and brought their involvement with the tribe to a sudden end.

On one of their many trips to get resupplied they were denied re-entry. It crushed them that, not only could they not return, but they were even denied the chance of saying their goodbyes and praying one last time for the tribe who had become their family.

The only comfort was that they had, by God's grace, been able to teach the tribe a solid biblical foundation. The villagers could now read and had some portions of scripture in their language. They knew their Savior and had come out of the darkness and into the light. Hallelujah, what a Savior.

When the political turmoil subsided, violence of a different kind struck the region. Mount Pinatubo, a once dormant volcano in a chain of volcanoes on the western side of Luzon, awoke from its four hundred years of slumber when a massive earthquake rocked the region in 1990. Baguio City, eighty kilometers away from the mountain, was devastated by the quake, which experts say triggered the horror the mountain would soon unleash.

In March of 1991, a succession of earthquakes hammered the region, which led to the first eruption of magma from the volcano. On June 15, the thing the Philippines had long feared finally occurred: Pinatubo erupted.
The volcanic explosion devastated the region, shooting ash more than thirty kilometers into the sky. The region was then hit by a massive typhoon, which pounded the whole area just north of Pinatubo. The devastation was immense.

Scientists reported that ash fell as far as Vietnam, Malaysia, and Cambodia. In my home country of New Zealand, the fallout from the volcanic eruption affected weather patterns for the following two years! And in the Luzon region, ash covered a radius of more than one hundred kilometers, (approx. 62 miles) killing trees, plants, and all other vegetation. Metro Manila itself, more than one hundred kilometers away, was doused in more than an inch of ash.

Nearer to the site of the eruption, the damage was most severe. Villages were buried in more than ten feet of hot ash. The mountainous terrain in the region was totally changed. Valleys were filled with ash, bridges were wiped out, and the once lush forest became unrecognizable. The weight of the ash collapsed the roofs of the buildings at Clark Air Force Base, damaging and destroying planes and other equipment. All of its personnel evacuated the area and never returned.

We lost all contact with the Wuh tribe after Pinatubo unleashed its horrors. The whole area up to where the tribe lived, on its quiet, hidden plateau was impassable. So remote and small, we believe the whole tribe we had come to love was wiped out in one horrific moment by the volcano—ushered out of this life the same way they were brought into God's kingdom: together as a tribe, in a single moment.

Chapter 19: The God of All Circumstances

About a year after first visiting the Wuh tribe, I was trying to get out of the southern part of the Philippines. The experience, like others before and after it, really tested my faith. I had to come to terms with Paul's command to rejoice in God at all times.

I was trying to get out of Davao to do some work in the northern part of Indonesia before I returned to New Zealand for a break. I had been working in Asia for over three years already, and was planning a trip home. I had a whole schedule arranged, and had prepared to meet a series of Indonesian churches. There were no commercial flights to Sulawesi, Indonesia from Davao in the south of the Philippines at the time. One could only book a chartered flight with a company that flew to Sulawesi on an irregular basis with its employees who worked in the foresting industry there. So I went to the company and asked if I could get on one of their flights.

They told me I could if and when there was room. The employee I talked to said, “Keep your bags packed. We'll give you a call when there's space on a flight, it'll be short notice, you might have just an hour to get here to the airstrip before the plane takes off.”

It was my only option, so I agreed, although it was pretty radical for me to have to just wait around in a state of readiness, hoping for a phone call to go. This was not the era of the cell phones and texting, it wasn't as easy then as it is now to communicate! A few days later, the telephone rang, and the airline told me to get out to the airstrip because there was a flight. So, I got my things and went.

When I arrived they apologized. The flight was now completely full.

I had been pastoring a small church in the area for a while. They had been without a pastor, and asked if I might fill in for the remaining time I'd be in the area. It had been a very effective time. We'd won many people to Christ and the church was excited and growing. I had become a spiritual father to the believers in the church, and there was a lot of tenderness toward me. They didn't want me to leave. The night before my flight to leave the region for good, the church threw a farewell dinner for me. It was an emotional night with lots of hugs and tears. Everyone was crying and saying their goodbyes.

When the airline told me they had to send more workers than expected to Indonesia on that flight, my heart dropped. I didn't like the idea of returning to the church. They had gone to great lengths to send me off with a big party, and we had all said goodbye. It would be awkward. I didn't want to have to go through all the emotions of saying goodbye again. I pulled out my wallet to see if I could afford to stay at a place near the airport. I had absolutely no money! I had no option but to return to the little church. For the life of me, I didn't want to, but I had to die to my pride, bite the bullet and face the embarrassment.

I got a ride back to the church. As I approached, a few of the members saw me come up the road and they ran out toward me, shouting happily. "We told you that you would come back. We know you are supposed to stay with us!" They were elated, but as you might imagine, I sure didn't share their excitement.

I had to wait another three or four days until the phone rang again, but finally I got the call from the airline that a flight was available. So I said my goodbyes all over again, and returned to the airstrip. When I arrived, the plane was in the air, circling the airport. We were still under Martial Law in the Philippines at that time, and all airports were heavily controlled. The incoming flight, waiting to land and off-load its passengers, was the one I was supposed to depart on.

The plane had not gained clearance, so the air traffic controllers would not give the aircraft permission to land. I stood and watched the plane circle for some time before it was sent back to Indonesia without landing. This has to be some terrible joke, I thought. I could see myself returning to the church again and the members excitedly running out to meet me yelling, "He's back, he's back!" Oh Jesus please, I prayed under my breath, this is nearly more than I can handle!

The third time the airline called, I went out again. Once I got to the airport, the agent frowned and started to apologize. "Sorry, sir, we had to take the front seats out of the plane to load it with machinery for a job we're doing. There's no room for passengers."

"What?" I exclaimed. "You've gotta be joking!" But joking they weren't. "How many times is this going to keep on happening?" I asked. "You call me, I come, but only to be told, 'Sorry, you can't board?'"

I didn't attempt to mask my irritation. "Every time you call me, you send me back!" I said. The agent apologized for my many inconveniences but this did little to settle my frustration. I could see the church members in my mind's eye as I walked round the corner to their street running excitedly toward me again, yelling, "He's back, he's back!"

I returned to the church, and sure enough the believers came running out to welcome me again. It was the focus of a running joke. What seemed tragic to me was a wonderful comedy to my friends. Every time I got a call, they told me that they prayed I would stay. God seemed to be answering their prayers and not mine, because every time I got a phone call, I prayed that I would be able to leave. I had made commitments in Indonesia, and I felt terrible for missing them.

The fourth time I got a call the night before the flight. The employee told me to be at the airport in the morning. I asked the man on the phone if the captain of the flight was in town. I found out where the pilot was staying and went to his hotel. I told the captain my story, and asked him to help. The captain assured me, "Sir, you get to the airport early, I'll get you on the flight."

I thanked him profusely. With an air of newfound confidence I returned to the little church, and happily informed the members that I would definitely be leaving the next morning. They all laughed and said, "You'll be back."

"You don't get it," I told them. "I'm definitely leaving tomorrow. I spoke with the pilot!"

I made sure I was at the airport early the next morning, but was immediately suspicious of the fact that there was virtually no one there. There was one lady at a desk. She told me that the flight I was supposed to be on was already on the tarmac and lining up for takeoff. I was shocked.

"That can't be possible!" I said. "I was told the flight wasn't leaving until 10:00 AM, it's only 8:30 AM! What's going on?"

She looked at me and shrugged. "The company flies when it wants to fly. They don't run on a commercial schedule, they operate on their own schedule."

"But this isn't fair!" I pleaded. "The captain assured me only last night that he would guarantee me a seat on this flight!"

"I'm so sorry" she said, "There's nothing I can do."

I asked her to excuse me, and then quickly jumped on the luggage conveyer belt before she could respond. It spit me out onto the tarmac. I ran out to the plane and waved my arms frantically at the captain whom I had spoken to the night before. He stared down at me from the cockpit. The jet engines were blasting as I gestured that I wanted to be on the flight. He signaled that it was impossible. I gestured again by pointing my finger to my chest and then to the airplane smiling broadly and hoping for a positive response, but again he signaled No. My heart sank.

What can I do? I thought. My only option was to either sit on the tarmac at the front wheel of the airplane to keep it from going anywhere, or to return to the church with my tail between my legs one more time. After wrestling with my incredibly frustrated emotions for a while, God came to my rescue with some much needed common sense. I accepted defeat, returned to the terminal, and recovered my bags from the lady behind the desk. I noticed that she was still in shock from my disappearing-act on the conveyor belt. Dear Jesus, I groaned, not again!

Once again, I started the long walk back to the church in Davao. I had been so sure that this morning I would finally be on my way to Indonesia, that I told the church members who drove me to the airport not to stay and wait, even though they reminded me what had happened on the last few attempts to leave. I assured them I would fly. After all, I'd personally arranged it with the pilot! Needless to say, when I finally got to the arrivals area, no one was waiting for me. They were all gone.

The airport was about ten kilometers outside of Davao City and I had no money. I would have to walk all the way back with my two heavy suitcases. It was a terribly hot journey back to the city, walking along the unpaved dusty coral roads. Within a few minutes my clothes were drenched with sweat, sticking to my skin like wet cardboard.

I felt angry enough for steam to be coming out of my ears. I was angry at the pilot and at the company, humiliated to think about returning to the church, yet again, and having to eat humble pie. Most of all, I was angry at

God for not helping me to get out of the city and make my appointments in Indonesia. With all of these emotions brewing inside of me, I felt like a walking time bomb just waiting to explode. Then, in the middle of my anger I began to sense the still, quiet voice of the Holy Spirit whisper to my heart: Just praise Me, Kel.

The words didn't soothe me as you might imagine. I could hardly believe God would have the audacity to make such a suggestion. I felt even angrier. Just praise You? I thought, when I'm feeling furious after all I've been through?

"No!" I screamed out loud. If God didn't yet know I was an unhappy boy, then I wanted to make sure He understood it now. I was about four kilometers into my walk and still furious. I was not going to praise God. I was absolutely unwilling. But those three little words, Just praise Me, continued to circle in my mind, like a song that keeps running through your brain that you can't stop. I wanted to get the words out of my mind, so I could enjoy my self-pity and stew in my anger. After all, didn't I have a right to be mad at my circumstances and at God for the mess I was in? There was silence for a while between God and myself as I trudged along the pot-holed gravel road that was so generously covering me in dust.

It wasn't long, though, until I heard the gentle voice of the Holy Spirit repeat those words again. Just praise Me, Kel. No matter how entitled to my emotion I felt, I couldn't shake that voice urging me to worship. My heart was unwilling, but about one kilometer later, my heavy bags in my hands, all alone on a deserted car-less road, I decided I had to do something about this voice. So out there on that lonely road I drew a deep breath and shouted "Hallelujah" to God, rather angrily, hoping God would leave me alone now. It didn't help. There was no willingness in my words. I felt defeated, frustrated, and angry, all at the same time. I couldn't find any joy within me, and that made me feel even more miserable!

I tried again. "Hallelujah!" I shouted, not as angrily, but still with angst and frustration. Shouting wasn't working. No matter how many times I tried this, I knew it was no way to worship The Almighty God. One kilometer resisting God's voice was terrible enough. There was no way I could walk the remaining distance in this state.

I realized it was going to be a long and horrible walk if I didn't deal with my heart, so right then and there out on that road I dropped my bags and lifted my hands to Heaven and began to pray, "Lord, I give You my anger, my self pity and my frustration. Please help me to deal with my hard heart. I don't want to continue like this, so I'm choosing to praise You right now, as You deserve." I began to sing quietly under my breath, "I worship You. Take my anger from me, please." I stood there for a while allowing God to work in my heart, the anger started to leak out of my pores, along with the sweat.

I picked up my bags again. They weren't any lighter, but I had done what God had asked me to do, and my spirit started to feel lighter for it.

I finally made it back to the church, several hours later! Sure enough, some of the members saw me come around the corner and the familiar words rang out, "He's back, he's back!" They ran up to me in great jubilation, excited to welcome me back. They had become convinced that this time I was really gone. Yet miraculously, in their midst, here I was again. They grabbed my dusty bags and joined me as we completed the last few steps along the street back to the little church together. "Just park my suitcases in the next room," I said. "I need a moment alone to ask God why I'm still here." Had I been more humble and spiritual, such an obvious question should have asked long before that moment!

Pastor Condoli

I went up to the little room in the church where I liked to pray. Before the words, Why am I still here, were off my tongue, a picture of a local pastor came to my mind. He was a man whom I had once met at a pastor's conference some months earlier in the city, but I couldn't recall his name. I could see his face very clearly in my mind's eye. I described the man's face to the church members and asked if they knew his name.

"That's Pastor Condoli!" one of the brothers said after hearing my description. I recognized the name as soon as I heard it.

"I have to find him," I said. "I believe the fact that I'm still here has something to do with him."

One of the members knew where he lived. They put me on the back of a motorbike and took me out to the outskirts of the city right away. I was going to pay him a visit. He happened to be in his back yard pruning fruit trees. He recognized me as I hopped off the back of the motorbike, and came straight across the yard to greet me. He was happy to see me, but I could tell right away that there was something terribly wrong.

I asked him how he was doing. He responded with a whispering, raspy voice that was barely audible. "Brother Kel, I have come down with a bad sickness. Every time I talk, I cough up blood. When I go to the bathroom, I pass blood. I cough all night long and keep my family awake. I can't keep any food down. The doctor's have put me on baby food, but I can't even keep that down, much less water." He looked at me with tears beginning to slip down his face. "I'm going to die! Who's going to take care of my wife and eight children? Who's going to look after my church? I've asked three other pastors to come, but they all have churches of their own." He started to weep. "My wife can't do the preaching because our denomination doesn't believe in women preachers."

I put my arm on the man's shoulder. I was deeply touched by his circumstances, and the huge effort it took for him just to speak. Wondering how I might help, I asked. "It's Sunday tomorrow. Would it help if I were to preach for you in your morning service? We could start with that, and then see how God leads us?"

He agreed. The next morning, when I arrived at his church, I told Pastor Condoli that the Holy Spirit had prompted me to speak on a subject I knew his denomination didn't agree with: divine healing. "I believe the Lord has laid it on my heart to pray for you today and speak on the topic of divine healing, would this be okay?"

"Just go ahead and do whatever God tells you," Pastor Condoli said.

That morning I gave a healing message. Several of the older ladies in the front of the church had rheumatoid arthritis. The Lord touched them so they were able to straighten out their fingers. Much of the church was in a state of surprise. The Holy Spirit was in their midst. There was power and energy in the church, they could feel it!

Pastor Condoli, however, didn't come forward for prayer. So, I told the gathered believers we needed to pray for him. I asked everyone to stand up and hold hands— something the church never did. The congregation was nervous, as some were reluctant to do such a thing. The church was very "proper" and such things were just not done.

"I know this is not common," I said, "but we're going to hold hands to show that we are united in our desire to see the pastor healed. Is that okay?" They finally agreed. They joined hands and I prayed a simple prayer, asking God to heal their pastor. At the end of the service, Pastor Condoli ran up to me and said, "Brother, surely God has healed me!"

"Well hallelujah, what makes you feel confident you're healed?" I asked.

"When you all prayed, it felt like something came undone inside my stomach, the way you'd pull a bow, to untie a wrapped gift." He stood beside me, holding his stomach and shaking his head, amazed. "It just came undone!" he said again.

"Wonderful," I replied.

It was their practice at the church to take out visiting pastors for lunch and so he took me to a restaurant. I ordered my food and when it arrived the pastor simply sat there with nothing in front of him. Somewhat mystified I said, "Well, pastor, you better order something, because I'm not saying grace until your food arrives."

"They don't have baby food," he said. "I can't eat."

"If you're healed you can eat anything," I assured him. "And while you're at it, order yourself a Coke".

"Come on now, there's no way!" he replied. "Pouring that down my throat would be like pouring acid on a wound, the worst thing in the world for anyone with a condition like mine. I can't even manage water, let alone solid food or Coke!"

"Listen, if you're healed you're healed," I said. "Now hurry up and order your food."

"Do you really think so?" he asked.

"Absolutely," I responded, and sat back in my chair. He made his order and his food arrived at our table. After giving thanks for the food, I noticed Pastor Condoli was only taking very small snippets of food to be sure he could swallow without pain.

"How does it feel?" I asked.

"So far so good." he said.

"What about the bottle of Coke," I queried. "You can't just sip that you know. If you're going to drink it, just pick the bottle up and drink the whole thing down in one shot."

"Are you crazy?" he asked with a strange look on his face. "The acid would burn my insides! It's gotta be the worst thing for me to drink."

"Not if you're healed," I reminded him. "Remember you just told me you believed God had healed you? Well here's your opportunity to know for sure. Just go for it and drink the whole thing down in one shot. We'll either have for ourselves a great miracle or a great explosion!"

So without any further encouragement from me, he picked up the bottle and said, "Here goes," and drank the bottle down in one large gulp. I smiled when he finished the gulp, and started to eat my meal. After sitting in his seat for a moment, waiting to see what would happen, he jumped up and kicked his chair back and started shouting, "I'm healed! I'm healed!"

People in the restaurant turned and gave us funny looks. Unaware of the commotion he was causing he happily continued to shout. "Sorry brother Kel, I have to leave right now and tell my family." So he jumped in a taxi and was off, to tell his family about his miracle. I watched him run off, thrilled to see his faith rewarded, and then he was gone. I was abandoned there on my own, with two full meals in front of me.

That night, I received another phone call from the airline. There was a flight in the morning and they assured me that I would get on the plane. I flew to Indonesia that very next morning. I now understood that the reason God had kept me there had been fulfilled in the healing of Pastor Condoli.

Pastor Condoli heard that my flight was leaving, so he met me at the airport to wish me well. "I have only been preaching half the gospel," he said, "but now I'm going to preach the whole message. I'll probably get kicked out of my denomination, but that's a risk I'm willing to take!" Even though his church began to grow in leaps and bounds, and quickly became one of the most successful in the denomination, one year later he wrote me to say, "We have been put out of the denomination for preaching the whole gospel. But I'm happy to inform you brother, all the members are in complete agreement with the preaching so we continue to grow in numbers and rejoice in His goodness."

Chapter 20: More of the Miraculous in Indonesia

When I arrived in Sulawesi, Indonesia, I had missed almost all of my appointments in the northern region of the country where I had been scheduled to minister. I apologized to my contacts for arriving three weeks late. Having had no phone numbers for any of them, communication had been zilch. They had no clue where I was. Thankfully, they understood and were not upset.

A new schedule was quickly set up, and before long it was time to travel to my next port of call, Surabaya. On the flight, I was seated beside a young man. We started talking, and I learned during the course of our conversation that his father was a pastor of a Seventh Day Adventist Church. He told me he would be picked up at the airport by a chauffeur, and asked if I wanted a ride. It sounded better than the bus, so I agreed.

We drove off to his father's place like a pair of princes. When we arrived, his father met us at the door and put his finger to his mouth indicating we should be quiet. "Who's this?" he whispered sternly to his son. We stepped away from the door and he began to interrogate me with a lot of questions.

I told him my name and who I worked for.

He looked puzzled. "Youth with a vision?"

"No, sir. Youth With A Mission," I repeated. Even though I repeated the name three or four times, he just couldn't seem to get it.

"Well, what is it that you actually do?" he asked me.

"We evangelize, preach the gospel and pray for people." Then I mentioned the story of Pastor Condoli who was just healed in the Philippines.

"You pray for people and they get healed?" he asked.

"Well this time they did," I said.

He thought about this for a moment. Then quietly he said, "My wife is very sick in the next room. That's why I gestured for you guys to be quiet when you first arrived. I didn't mean to be unfriendly. I was just concerned for her. Would you be willing to pray for my wife?"

I told him I would love to. I went with him into the room where she lay in bed, and said a simple prayer. Knowing Jesus is the healer, I prayed a few words and asked that his healing would touch the man's sick wife. The man was so thankful he drove me to the train station even though it was only a five-minute walk from the house. He even bought my round trip train ticket to Jakarta, and told me to leave my heavy main bag with him, saying that he would store it at his house while I was away so I could travel lightly on the overcrowded train.

A week later I returned to the city of Surabaya after my meetings in Jakarta. I walked up the road toward his home, and found him standing outside. He knew the arrival time of my train and had been waiting for me. When he saw me he immediately yelled out, "Brother Kel, brother Kel!" Then he ran toward me in great excitement. "When I took you to the station last week, before I even arrived back home, my wife was already up out of bed, healed!" When he caught his breath he said, "You've got to come to our churches and teach about healing. No one knows about it in our denomination. I'll open the whole of East Java to you to preach about divine healing!"

I looked at him amazed. Not that his wife was healed. And not that God was opening doors, but because I had suddenly realized just how important it is to be on God's schedule at all times.

Missing appointments that I thought were important, because of my repeated failure to catch a flight in Davao, seemed like a huge disappointment to me. Instead, these were divinely orchestrated events God had used to direct my steps.

His ways are not our ways. Sometimes God's schedule can be in a totally different ballpark than ours. If I had got my way, I would not have seen those miracles, nor would I have had the opportunity to reach all the people I was able to minister to through the healing of Pastor Condoli and the Seventh Day Adventist Pastor's wife. Without God's help, I might still be in the Philippines, trying to catch a plane!

These experiences taught me a crucial lesson. I've tried ever since to live with enthusiasm towards my present circumstances, responding to the quiet voice that says, Just worship Me. It's not always an easy thing to do, but gratefulness to God and worship in our present circumstances, especially when things aren't going right, keeps us in step with Him, it can open the resources of Heaven for us to do effective work.

Chapter 21: Confirmation of My Calling

By this time, I had been out of New Zealand for three and a half years, and planned to head home for a much needed break to seek God concerning my future. In the summer of 1973, I informed Ross and Margaret of my plans, and they came up north from Mindanao to take over as leaders of the work the Lord had started in Baguio.

As I was about to leave for New Zealand, Loren Cunningham called together all of YWAM's leaders worldwide for a meeting in Osaka, Japan. YWAM was about to purchase a ship called the Maori, an inter-island ferry used between the north and south islands of New Zealand. It was a crucial moment in the life of the mission organization, and in my life as well.

The conference was an intense, soul-searching time. Ninety-three leaders from around the world met to touch base, seek guidance from God, and to pray in the sparsely decorated conference room of an Osaka hotel. After one of the sessions, Loren was walking by me on a footpath.

"Kel, what are you going to do after you return to New Zealand?" he asked, aware of my plans for a furlough.

"I'm not sure," I said, which were the only words I knew to say. My time in New Zealand was an interim between the amazing years I just spent in Asia and the next stage in my life. And that's what I told Loren.

Loren responded, "There are no interims, Kel. You're either in or you're out." He looked at me for a moment, "you need to hear from the Lord and make an announcement at this conference about what you're going to do next."

I was a bit flabbergasted, it was Friday afternoon and the conference was set to conclude on Saturday morning. Lord have mercy, I thought to myself, that's coming up real fast since Saturday morning is only hours away! I was to announce to the group what my future plans were by then? I walked to my room and shut out the world, sat on my bed, and began to pray.

I started to go through the prayer principles I had learned from Joy Dawson years earlier when I first set out to do mission work. I made sure my heart was clean before God, so that I could hear from him anything He wanted to say. I invited God to speak to me, and asked him to silence the voice of the enemy. Then I did my best to hear God speak to me about what I was to do in the future.

I sat in the room for quite some time. I looked at the clock and it was past two AM. Already it was the wee hours of the morning and I hadn't heard a thing! I finally became so frustrated that I told God I had tried everything I learned about hearing His voice but it was to no avail. I felt as though I had been tangled up for the last few hours in a spiritual jungle gym, trying everything I knew about prayer with no results. In sheer frustration, I determined to myself, If God didn't speak to me, then come tomorrow morning I'll simply have nothing to say. I finally became so exhausted in my desperate struggle to hear from God that I gave up and went to sleep.

When I woke just a few hours later, I didn't have peace in my heart. It was weighed by the burden I felt of having to announce my future plans to the group. In a few hours, I was supposed to communicate to all my friends and fellow workers what the future had in store, but I awoke without a single clue. I imagined standing up in the room of people with nothing to say, so I resolved in my heart to simply announce, "I don't know what I'm going to do after I return to New Zealand." That's all I could truthfully say at the time, however it pained me.

I thought over the previous night's struggle to hear from God. I did my best to hear from him, but couldn't hear a thing. I didn't know what else to do. In my heart, I decided to give up. As the dust of my honest pronouncement of giving up settled to the floor, a gentle voice, like a soft wind, tugged at my heart. Well, now that you've stopped trying so hard, I can say something.

And just as quickly, three questions, like arrows through a banana leaf, shot through my mind. Did I call you to missions? Instantly I answered, "Yes."

Did I tell you to return back to New Zealand? My pastor had written suggesting I needed to return and reacquaint myself with the fellowship, so I said Yes again.

The third question had a different response. Did I tell you to stay in New Zealand? I thought for a moment, but without hesitation said, "No."

This whole conversation unfolded between my mind and my heart in a moment. Moments earlier, my heart felt so troubled, until I experienced a peace that passed all understanding. Well, what's the problem then? God asked.

There was no problem! I suddenly realized there was nothing to fear. "I'm supposed to go back into missions, that's what the call of God is on my life," I found myself saying aloud, as life and excitement coursed through my veins.

With a thankful heart, then, I was able to stand in the very conference room I had once dreaded. Joyfully I shared my plan to go back into missions after my time in New Zealand. It was a crucial moment in the path of my life, a moment in which, through Loren's encouragement and God's knowing grace, I was able to put a stake in the ground. This became a signpost I could look to in the future to identify the call of God on my life, one that would mark out the pathway I would walk for the rest of my days.

I say the experience highlighted God's knowing grace because when I returned to New Zealand, almost as soon as I arrived, I started to receive invitations from various churches to join their pastoral staff. But with each offer I was able to graciously and humbly decline, saying, "Thank you very much, but God's call on my life is to Missions and I'll be returning to the mission field very soon."

I didn't have to pray or agonize over a single one of those invitations because they all required that I stay back and reside in New Zealand, which God had clearly revealed to me in Osaka was not his call on my life. Although it was a frustrating time in Japan, the difficulty was brief and the clarity with which God spoke to me spared me from a great deal of confusion during my furlough in New Zealand. Instead, I was able to enjoy my time at home before I set off for my next adventure.

Chapter 22: Micronesia

It was wonderful and refreshing to be back in New Zealand after three and a half years abroad. I enjoyed how orderly everything was and the fabulous eye-catching scenery. I headed outdoors for some good runs and enjoyed it all, and most of all, reconnected with friends and family.

"So, Kel, are you here to stay?" asked people in my church, sometimes mischievously.

"Not for long," would be my reply, but I appreciated that I was welcome and wanted at home. While in New Zealand I learned that Loren was asking all YWAM leaders to come and do a school in Hawaii. As was often the case throughout my life as a YWAMer, when Loren communicated, my ears perked up and I listened.

I told my pastor about my plans to attend the school at Loren's request. "The man's a modern day apostle, isn't he Kel?"

"Sure is," I said. "God really has used him to pioneer short-term missions in the world."

"What is it about YWAM that you love so much?" asked my pastor. "What motivates you to travel all over the world?"

"You know, from the time I was young I had a vision in my heart to be a missionary. And all my life, I've felt comfortable in interdenominational churches. YWAM is just a perfect fit for me as they're an interdenominational mission. I also love the fact that if you have a vision to go anywhere in the world, YWAM says 'God bless you, go for it!'"

"But doesn't that make it possible for young people to make a lot of unnecessary mistakes, because of their inexperience?" asked my pastor.

"Loren is taking a certain risk on all of us young leaders," I agreed. I thought of my early experience in the Philippines, and knew it was true. "But I think he believes that harnessing the passion of a young man to do something for God is more powerful than the fear or possibility that he might fail."

My pastor thought about this for a while. "I guess the victories young people like you can gain in God's name are worth any of the mistakes you might make in the process. Making mistakes is better than having no young people in missions at all. I'm glad you've found a place where you belong, Kel."

"Me too."

Soon after, I packed my bags to attend the School of Evangelism (SOE). As YWAM's training progressed we began a six-month, entry-level school required to become a missionary with YWAM called a Discipleship Training School (DTS). It was 1974 and YWAM was still in its infancy. I, and many other YWAM leaders, had worked for years in the mission field as YWAMers, but we all agreed it was important to connect in Hawaii for the newly formalized training. We believed it would be an excellent way for everyone to come to the same understanding of the mission's values. Today, anyone wanting to serve with YWAM begins by attending a DTS just like the one eighty or so of us early leaders attended in 1974.

The school was very powerful. YWAM leaders from all around the world who were working as missionaries met in Hilo, Hawaii. Most would return to the spots on the globe where they already had established work. Some had come, like me, to be equipped and then launched to a totally new place in the world. The school deeply impacted me as a young man. We learned heaps of valuable truths that we've continued to live and teach throughout our lives. I was especially impacted by teaching on Spiritual Warfare, the Nature and Character of God, and the Work and Power of the Holy Spirit.

As the school neared its end, we started to talk about outreach, discussing when we left the school, how we planned to put what we learned into practice. Early in the school I had my eye on Brazil. I'd never been to South America, and the country known for the Amazon rainforest with its rich wildlife, natural beauty, and many unreached tribal groups intrigued me. I discovered, though, that Jim and Jan Stier had already volunteered to pioneer work on the continent. So, I decided to head to Indonesia, remembering my time in the country with fondness.

Some of the other YWAM leaders approached me when they heard of my plans. "Kel, we just found out there's a South Pacific Games in Guam next year."

"What are the South Pacific Games?" I asked, always interested in the possibility of connecting with athletes.

"It's a sporting event for all the South Pacific countries," said my friend Dean Sherman. "We heard you want to head to Indonesia. You could take a detour."

"It's kind of on the way," I laughed.

I prayed about it and felt this was of the Lord. I invited another young man, Steve Shamblin, from North Carolina to join me. Steve was also attending the SOE. We set out first traveling all through the islands as a trailblazing trip. We visited the Marshall Islands, the Kiribati Islands, Pilau (Balau), Yap, and Guam. The islands we traveled to appeared on the horizon like jewels. White sandy beaches glistened in the emerald green water. What a beautiful part of the world!

Arriving and taking off from the Islands unsealed coral airstrips would send a rush of adrenalin through anyone's veins. Arriving planes approached low over the ocean, touching down on air strips only a matter of feet above sea level. Instantly, when the plane touched down, the pilot would apply reverse thrust with his foot hard on the brakes to bring the plane to a stop before hitting the end of a perilously short runway.

On departure, the pilot would taxi the plane to the very end of the runway before turning it around. To us wide-eyed passengers, it looked as though the nose of the plane was over the edge of the cliff! Soon as we were ready for take off, the pilot would give full throttle and the jet engines would roar into life. The blast of the engines would set you back in your seats and send clouds of coral dust hurtling into the air, temporarily blotting out all vision.

The plane would start screaming down the runway, then all of a sudden the air would clear and once again you could see the end of the runway looming up. A scary feeling started to come over you as the jet hurtled towards take off.

There was no room for pilot error. The end of the runway came abruptly with a sheer cliff that suddenly dropped off into the deep blue waters of the Pacific Ocean. All along its length, signs on the edge of the runway counted the meters towards the end. Before you knew it, you'd see the last marker clearly in view before the pilot, at the last possible second, tilted the wings for lift off. Oh what a relief it was to be airborne!

Other than flying in and out, life on the islands was laid back. The people we met were pleasant, quick to smile, yet still shy. We traveled around, trying to learn what the needs of the people were, and what ministries were represented.

Christianity had come into the islands in the 1800's. The churches had meshed into the fabric of the society, and there was a distinct sense among existing churches that we YWAMers were outsiders and that the islands belonged to the churches already established. We were aware there might be some resistance to YWAM coming into the Islands from the get-go.

When Christians had first arrived on the islands, they had a significant impact. Now, instead of there being a living faith, some of the islands had simply become religious. We soon recognized that there were many people on the islands who weren't attracted to religion, especially the young people. Church, like in many other parts of the world, represented a traditional part of the social structure, attendance was propped up by rules. Our desire was to show that there could be life, joy, and adventure not just inside the walls of a church, but outside of them, too. We had a message of hope to share with people all over the world!

The Games

Soon the South Pacific Games were upon us. The first Games were inaugurated in 1963, and ever since have taken place every four years, like the Olympics. Many posters announced the purpose of the Games: To create bonds of kindred friendship and brotherhood amongst people of the countries of the Pacific region through sporting exchange, without any distinctions as to race, religion, or politics. Like any inter-Island event, these Games were filled with excitement and invigorated by the energy and liveliness of youth.

Years earlier, I had dreamed of competing for my country as a runner. Now I was pleased to be there as someone who introduced athletes to Christ. My athletic energies had been re-channeled to reach others for God, and as we geared up for the event, I found myself thinking, How many of these people, like me, can discover the Good News and have their energies refocused towards Christ?

Our ministry was to athletes. We wanted to show the international competitors that there was more to life than sports, and to show them what great feats they could accomplish for God if they dedicated their lives to Him in the same way they had dedicated their lives to their sport. We slept at the Wittengale School. During the day we would join up with Teen Challenge. They had a church on the island filled predominantly with young islanders, or Chamorros (which is what local islanders called themselves), and Filipinos. We worked closely with pastor John and Eva Penada who led the local believers. The whole church got involved in that outreach, together with young people from other nearby churches. We met for times of worship and training, and then headed out to the Games.

There were about one hundred participants in total. We met athletes where we could: during the events, at the athlete's village after their day of competition, and in common places. We simply started conversations with athletes, met them individually or in small groups and talked with them. We did a lot of guitar playing and singing to draw crowds, who, curious, would want to learn what we were singing and why. It gave us unique opportunities to share the gospel.

We didn't have a complicated strategy when we shared our faith. We were all just young and excited about life. I enjoyed getting to know athletes and talking about my life. I would say I had been an athlete myself. I would share about my training experiences, and then share that since I became fully committed to Christ I got on the winning team. "So you guys, if you want to be on a winning team after you finish the Games, I can introduce you to the Winner!" With thankfulness I can say many athletes at those Games joined the team that will win eternal glory.

Years later, when I was traveling through Vanuatu, an island nation in the South Pacific, I met the Superintendent of the Assembly of God church there. I was on the island to help with preparations for the Anastasis, YWAM's mercy ship, to arrive at the island. As we got to know each other, I asked him how he had come to know about Jesus and he said: "I was on a boxing team at the South Pacific Games in Guam." I pointed to myself and told him I headed up that outreach. We had a good laugh. The whole boxing team, in fact, accepted Christ as their Savior during the Games!

As I traveled throughout the islands in the years that followed, I met numerous people in ministry who were saved at those Games. Our ministry at the sporting event had a significant effect in the region. What a thrill it was for me to see that our work was fruitful.

Chapter 23: A Short Term Trip Leads to Long Term Missions

Many of the young people we befriended during the Games wanted us to stay on the island of Guam. They asked us to remain there after the event because they loved YWAM's teaching and felt at home with it.

The idea seemed a good one to me. In turn, I enjoyed their company and was happy to stay longer in Guam. So, I prayed about it and soon felt God encouraging me to establish YWAM in the Islands. I committed myself to the Micronesian Islands for several years; first however, I had to sort out my travel visa so I could stay in the region long term. This required that I return to New Zealand to sort out the paperwork, a process that took nine months!

Steve stayed in the islands while I was working out my visa. Shortly before I was due back, however, he needed to return to North Carolina. So, when I arrived in Micronesia, it was to pioneer the work on my own. By this time it was late 1976. I had a lot of success among the young people of the island of Guam because they were eager to be discipled and to learn all they could about God. One of our endeavors was to mobilize many of the local churches to prevent gambling from being established on the islands. Our joint efforts were a success.

Saipan

By 1978 the time was right to expand the work further in Micronesia, so a small team of YWAMers and myself set our eyes on Saipan. Saipan is the largest island in what is known as the Commonwealth of the Northern Mariana Islands, which has been an unincorporated territory of the United States since 1986. It served as a good place for us to develop the mission, as well as a good launching place for mission trips into the other islands. In order to move and establish the work more effectively, though, I needed a bigger team.

Loren came through the islands and visited Guam at the end of 1978. As I started to talk with him about developing YWAM in the region, I said, "Loren, I need to raise up a team here if the Lord wants me to stay in Micronesia."

Loren thought about this, and then told me about a Discipleship Training School that was running out of King's Mansion, a YWAM property in Kona, Hawaii. "Why don't you come out to the Big Island?" he asked me. "Dave Gustaveson is about to lead a school there. He could use some staff. You could help with the school and when it's time for outreach, you could bring a team to Saipan to pioneer the work there."

It was a great idea! So off to Hawaii I went for a three-month stint to staff the lecture phase of the DTS. There were sixty students in the school, and the group split up in teams to go to different parts of the world. I arrived in Saipan for outreach with a team of eleven students in 1979.

A lot of people at the time believed missionaries spent their entire lives in one place among one group of people. So what Loren encouraged me to do by bringing a team in for a short period of time was rather revolutionary. For a long time, mainline, established missions organizations took issue with short-term missions; short trips weren't encouraged. For years, at mission's conferences, short-term missions were put down and ill regarded.

At one particular conference, when the inevitable topic was raised, I was given an opportunity to share what a huge blessing short term teams had been in YWAM's pioneering efforts, both in the Philippines and in Micronesia. Today, it's seen as part of the normal growth of missions. When God is doing something new, the established status quo often fights the new thing. Over a relatively short period of time, though, God has used Loren to pioneer short-term missions in the world. The results have been phenomenal.

Even though the outreach was scheduled to be two months, all of the students extended their time and stayed for twelve months. Through that experience, a number of the students on that trip became career missionaries.

Without the help of those eleven students, I would not have been able to pioneer the work in Saipan. Their faith and energy helped to found ministries on Saipan and various other Micronesian Islands that continue to this day. Short-term trips are a fantastic way to get your feet wet, as they say, in the mission field.

Over the years we ministered in Saipan, our team grew from eleven to over forty workers. One of those students was Matt Rawlins. He worked in Saipan alongside me until I left for the Philippines a number of years later.

Nelson and Diane Elliott, a Kiwi couple, came with me to the island along with a small team of Guamanians. Keith and Monica Winston, and later Mark and Laura McPolly, joined us too. We purchased a twenty-five-year lease on a property in Capitol Hill, a strategic location in the Mariana Islands that was first established by the Americans in the late 1940s.

Chapter 24: A Shark Story

Saipan is truly beautiful. On the eastern side of the islands are clean white sandy beaches; offshore is a coral reef that forms a stunning lagoon. The western side of the island is totally different. Jagged, rocky cliffs burst from the ocean's reef and stretch more than a thousand feet, revealing a landscape that is totally wild, rugged, and beautiful.

Early on during my time in Saipan, after spending a day praying and dedicating to God our new coffee house as well as another ministry building, I decided to go with my friend Craig Thorne on a scuba trip. It was a balmy day at the end of March, perfect for a dive. There's a world-class diving spot that is not for the faint of heart, called the Grotto. The Grotto is an inland cave, a hole in the earth where the sea comes inland about forty feet underneath the cliff. It is accessed from the road above the cliff by a steep stairway built into the rock wall. This magnificent pit is accessible to locals and tourists alike. It is a relatively safe entry point to begin an ocean dive.

There are numerous underwater caves and inlets along the eastern coast, and most of them look identical. Divers who make their way out to the ocean and wish to return to the Grotto entrance must know precisely where the underwater entrance is. Otherwise, divers will get lost in the underwater tunnels looking for the correct route to the surface. Because scuba tanks hold limited reserves of air, quite a few divers have entered the wrong cave, gotten lost, run out of air and drowned before they could ever find the correct path back to the Grotto.

To do this complex dive safely, it is important to go with someone experienced, someone who has done the dive before. One of Craig's friends was perceived to be just the right man. Or, so we thought! His name was Dan, he told us he had done this dive six times previously, so we were both confident that we could do the dive with ease in his company.

We walked down to the Grotto along with some of our friends who planned to pick us up after an hour or so. Scuba equipment in hand, we descended the steep stairs in the warm tropical air. The sky was cloudy, and there was a strong breeze blowing in off the sea that day. A massive rocky outcrop separated the Grotto from the ocean, if you climbed it, you could watch the water slam onto the outside walls of the Grotto and against the cliffs that went on for miles.

We entered the pool and noticed that the water was surging. We didn't realize that once we were in the open sea this would mean great danger. Excited about the dive, we did our final safety check, making sure our equipment was in good order, and that the tanks were full of air. Again, we discussed what we would do on the dive. We had about an hour of air available for our adventure.

We waved goodbye to our friends and slipped under the water. Dan was the leader of the dive, immediately he motioned for us to begin our descent. We followed him down some fifteen feet and swam under the cliff until we reached the ocean. The water was warm, no one ever wore wet suits, and it was clear. Going through the cave was somewhat un-nerving for this first timer. I didn't like the claustrophobic feeling of being in a dark tunnel, it was a bit freaky. I didn't like being close to the walls or the roof, so I focused on the light at the other end. As we exited the Grotto, we were astonished at what lay beyond.

If you've ever been snorkeling or scuba diving before, then you know that the ocean is a stunning and beautiful underworld. Three caves split in different directions, and through them we could see the inky deep blue of the ocean beyond. It was Craig's first dive on the island of Saipan, and he was set on seeing the two things he had yet to see while diving in the Mariana Islands: fan coral and sharks.

It was an eerie feeling to be diving into a completely silent world—silent that is, except for the sound of used air being ejected from the breathing apparatuses. Soon we arrived at the edge of the ocean wall. There before us was the vast Mariana Trench, the deepest ocean trench in the world, over thirty-five thousand feet deep at its deepest point.

The ocean floor dropped away immediately and the temperature of the water began to change as we dove down. We were accompanied by schools of fish, and surrounded by thriving coral and anemones. We had entered an extravagant world of beauty. We swam for about fifteen minutes when Dan suddenly stopped. He swam up to Craig's depth and pulled at him indicating that he needed to check his pressure gauge. Through his mask I could see he was anxious.

Dan made a quick about-turn and started swimming in the direction of the cave where we had come from. Because he was the dive leader, and because one of the first rules of diving is that you always stick together, Craig and I followed suit. As we made our way back to the cave, Dan made another about turn, leaving the cliff line, and swam frantically for the open ocean. We followed again, and watched him surface. There was nothing we could do. We had to surface as well.

The water on the surface was rough. It was hard to communicate. When we pulled our mouthpieces out to talk, the waves crashed against our faces and we swallowed a lot of water.

"What's wrong?" Craig asked, tilting his head back, trying not to choke on the seawater.

"I don't think I've got enough air in my tank to make it back!" Dan shouted back.

"Impossible!" I yelled back. "We checked the tanks before we left."

By this time Dan started to panic. He grabbed and clutched at both Craig and I in the rough water and pushed us under. We had a terrible situation on our hands. Our supposed expert leader was panicking and insisting that he didn't have enough oxygen to make it back into the Grotto!

Not wanting to use our oxygen tanks for fear of sucking them dry, we had to think quickly. We were treading water, yelling to each other over the roar of the crashing waves. We now saw what had caused the surge when we first entered the Grotto: a very stormy sea. White water from the breaking waves frothed around us.

Over the roaring of the waves, Dan and Craig suggested that I attempt to make it to the Grotto and sound the alarm for help. This didn't seem like very wise advice to me at all, and I wasn't about to have any of it! I had never gone diving at the Grotto before, I had no confidence that I could identify the correct entrance and find my way back. It was a crazy idea. We had to come up with something better than that. We were at our wits' end and becoming overwhelmed with the impending danger.

"We have to go back now!" I said. If we had any chance of getting to the cave we would have to take it right then.

Dan continued shouting, "I don't have enough air to get back to the Grotto!"

We knew we couldn't swim for the jagged rocks around the face of the cliff. There was no mercy from the violence of the raging seas that relentlessly pounded those rocks. Besides, a diving school instructor had previously warned us that many divers had died that way—dashed to pieces trying to come in on the rocks. We were in a real bind.

The only possibility was to try and swim up the coast to Bird Island, which was over a mile or so away. Behind Bird Island was the only place that didn't have jagged rocks. There was a small beach. The swim promised to be a marathon effort in such heavy seas, but it was the only real possibility if we were to have any chance at staying alive.

I regularly swam out a half-mile from the shore to the edge of the reef. There was an old military tank which had been caught in a hole in the reef while being offloaded by the US navy during the Second World War. I never knew that those long swims on our days off were the necessary training I needed for what would now become a swim for my life.

Dan was so panic-stricken he was vomiting into the sea. He was completely terrified and couldn't move. He grabbed onto my scuba strap with an iron grip. Craig also held onto him by one hand.

I noticed some movement below my flippers as we treaded water. Reef sharks! Earlier Craig had expressed that he wanted to swim with sharks. He now had his wish, and I can't say I was happy for him. The sharks began swimming in circles beneath us, perhaps curious about these strange creatures in their territory, or perhaps drawn by the smell of Dan's vomiting. We really had to move!

One of the sharks started to swim in a half a circle, back and forth, all the while getting closer to us. Not a good sign, I thought to myself, I don't like the look of its behavior. I began to pray with every bit of faith that I had left inside me, and rebuked the sharks.

"In Jesus name, be gone!" I shouted in my heart.

The whole lot of them took off. Whether they just left on their own accord, or God prompted them because of my prayers, I don't know, but they left, and I heaved a huge sigh of relief. I could breath a little easier. Thank you Jesus!

With no other options before us, I started to breaststroke toward Bird Island. I pulled Dan through the rough sea like a tugboat through the water. He was literally paralyzed with fear. He couldn't swim and was sure he was going to die. He gripped my tank strap with one hand, and held on to Craig with the other, impeding him as well. I had turned my oxygen tank off, to conserve the air I had left, so I was using my snorkel to breathe. This made the going exceedingly difficult. The sea was stormy and turbulent with breakers crashing all around us, endlessly filling my snorkel with water.

I would blow out the water then frantically gasp for another quick breath before the snorkel filled back up. Many times, I wasn't quick enough and ended up with a mouth full of seawater. Half choking with each breath, I used every last bit of oxygen I had in my lungs to eject the water and desperately try and clear my snorkel before it filled with water again. It was like this for the entire swim. Our situation was grave, life threatening, and totally exhausting.

At one point, we found ourselves in the middle of a floating mass of jellyfish, which stung our exposed flesh. The fear of sharks and other creatures that we couldn't see below us was an ever-present thought on our minds. Dan clung vigorously to my oxygen strap and it began digging into my skin causing it to bleed. This was especially bad, since sharks can smell blood a mile away. Worse yet, we swam against an opposing current, which meant our progress was painfully slow.

It was the hardest swim of my life. My legs began to cramp, but we could not afford to rest or the current would drag us back. It was all out war; my legs screamed for relief, but my mind was not willing to let my legs rest and concede any of the hard-fought distance we had gained. We had to continue swimming at all costs—cramps or no cramps. Our pleasant day of scuba diving had turned into a dreadful nightmare.

We swam and swam and swam. The fatigue was so great, the only thing I could think of was to keep moving, to make it to the island before dark. Survival was the only thought on my mind. We had to make it! As we approached Bird Island, a huge sea turtle swam near to us and I thought, Oh Lord, You are wonderful, You've brought a large turtle that I can latch on to that will pull us to the island!

As if hearing my very thoughts, the turtle turned away from us and swam off. My heart dropped. I didn't have the strength to swim any longer. Right then, even as the turtle, which was my last hope, swam away, the current abruptly changed. We now found ourselves being drawn towards the shore.

We went out on the dive at 3:30 PM, and by the time we arrived on the mainland shores directly behind Bird Island it was getting close to 6:00 PM. In the end we were helped ashore by a powerful incoming current, but each time we tried to stand up, the waves repeatedly knocked us over. We climbed onto the beach like shipwrecked men, our legs were so shaky we fell onto the sand unable to walk, gasping for air, thanking God that He spared our lives. Never had sand felt so soft. Never had any beach seemed so beautiful.

After lying on the beach, collecting what energy we could, we finally stood up. Already the sun was dropping in the sky, and soon it would be dark. We faced another problem. We had to carry our heavy gear, climb over the sharp rocks of a steep cliff, and walk, barefoot, through thick jungle as we tried to find a road before darkness descended. After that death-defying swim, our gear felt like a thousand pounds in our hands.

The excitement didn't stop there. Soon, into our bare-footed trek, as we raced against time to make it to the road before it became too dark, Craig stopped in his tracks.

"Kel, what's that?" he asked.

I looked at the metal cylinder in front of us in the path.

"It's a live bomb shell." I said in shock.

In front of us lay a large, undetonated metal casing of an explosive device, a relic from the Battle of Saipan in World War II. We surely didn't want to do anything to risk triggering the explosive, so the three of us gingerly stepped around it. After about an hour of weary trudging, we found the road and flagged down a pick-up truck. A kind stranger drove us all the way back to the Grotto, where our concerned friends had been waiting, fearing the worst.

And just like that, our nightmare ended. I'll never forget that afternoon. God spared my life, along with the lives of the other two men. I knew there was more work He had for me to do.

Chapter 25: Pioneering YWAM in Saipan

After World War II, Saipan became quite dependent on the United States. The United States gave flour, sugar, and staple foods by the shipload, and distributed it to the inhabitants of the island. Even though The United States had the right intentions to support the local population in the years after the war, a consequence of these handouts was that the islanders become dependent on constant US aid. We discovered that some locals were getting far more flour, for example, than they needed. Instead of only accepting enough flour for the needs of their family, they would take heaps more and feed it to their pigs. American taxpayers were paying for this waste of good food.

We saw that these policies, in the long run, would not be beneficial for Chamorros or for Americans. So we purposed to start an initiative that would help the local population become self-sufficient again, rather than dependent on the supplies Americans continually brought into the island.

The Chamorros proved to be bright and resourceful, like their forefathers. Prior to World War II, the Japanese had occupied the islands of Micronesia. They had planted sugarcane over about two thirds of the island of Saipan. The Japanese required Chamorros to work in sugar plantations. An old Chamorro man told me that cultivating and harvesting the sugar cane was work they despised.

The men had come up with an ingenious plan. Though they were constantly under the watchful eye of the Japanese, they patiently waited for moments of opportunity when the soldiers were distracted. They would seize such moments to sabotage the machinery they used to process the sugar cane, breaking parts and throwing them into the bush until the machine wouldn't work. The Japanese had to wait long periods of time for replacement parts to arrive from Japan and in the mean time the locals would disappear back to their preferred lifestyle as fishermen. Sick and tired of the Chamorros' antics, the Japanese had decided it would be easier if they just worked the land themselves!

The Americans fully ousted the Japanese towards the end of the war. When the United States military entered the war, after the infamous assault on Pearl Harbor, they came to the South Pacific. The US used a chemical defoliant spray, similar to Agent Orange used in Vietnam. This killed the vegetation so that they could locate where the Japanese were hiding. The chemical stripped vegetation of leaves, and had a highly negative impact on the environment. By the end of the Battle of Saipan, nearly all the 30,000 occupying Japanese had been killed and the island was devastated by the war. Similarly, American troops captured Guam and freed it from the Japanese, who had brutality mistreated the islanders.

As the United States helped to rebuild the region, to stop erosion and nutrients from seeping from the soil, they flew over the islands and dropped tangan-tangan seed to replace the vegetation. It is a legume, rich in nitrogen, which returns nutrients to the soil. Tangan-tangan is a fine, frilly, fern-like greenery with sweet and plentiful leaves that animals like to graze on. The tangan-tangan thrived and now has become a jungle that covers the island.

When you plant something into the soil of the beautiful Micronesian islands, this ideal tropical climate enables the seed to grow and bear much fruit. Much of the work we did as YWAM in those first few years on Saipan was just that. We planted seeds that would later bear fruit that others would harvest for the glory of God. We spent much time in prayer, and did all we could to share the Good News with the islanders.

We led many to Christ and were running numerous home Bible studies across the Island. We ran a Coffee House and Christian Book store we called The Upper Room, we had the joy of baptizing our many converts in the ocean beside the Royal Taga Hotel. Evangelistic teams were sent to all the Micronesian Islands, where many were led to Christ. We were privileged to make a significant impact in Micronesia for the glory of God.

The work wasn't always easy, though, in the same way that working land covered in sugar cane or tangan-tangan under the hot tropical sun wasn't easy. Many people on the island were either resistant to religion, or were so steeped in their own religious traditions they did not want to respond to the message of the gospel. Still, we were diligent to witness and to pray for the islanders. We organized a torch run where we ran around the entire island carrying a torch similar to the Olympic flame, claiming the scripture that He would give us the land wherever our feet trod. All the while, we prayed and worked towards revival in Micronesia.

In years to come, we received reports that other ministries had been fruitful as a direct result of what we had established in the mid 1970‘s and early eighties.

One of the first things we did on Saipan was to start a micro farm. We hoped to establish a model farm to demonstrate that the island could be productive and that its inhabitants could become self-sufficient. We saw this as a means to help the islanders, to teach them a trade, to become self-providing, and to help them recover their own dignity and self worth as a people. Establishing a micro farm was not a simple process. We had our work cut out for us.

Plagues of Biblical Proportions

In all ministries there will be opposition. One of our biggest foes during that time was a small, shelled creature: the African Snail. There were thousands upon thousands of them! The Japanese, fond of eating the snails, brought the invertebrate crustaceans wherever they went throughout the Micronesian Islands to supplement their diet. By 1941, the Japanese had their strongest presence on the islands as they hurtled forward to war. Some thirty thousand soldiers bunkered down on the island. They set up a base, developed the fishing and sugar industries, raised snails, and looked toward the West with thoughts of war.

All those years later, after the bombs had dropped and the world had reached its treaties of peace—long after Saipan and the other nations of the Commonwealth of the Northern Mariana Islands had been loosed from the Japanese grip by the United States—we had a new battle to fight. This was the infamous war against the snails.

These pests could grow to more than seven inches long. Overnight, they could wipe out any crop we planted: string beans, okra—you name it. Sprouting vegetables would be shorn off to ground level if a mass of snails migrated into the area. And migrate they would. Tens of thousands at a time would merge in a long slimy line many feet wide and travel together.

When they approached our farm, they literally covered the road. If we happened to be driving up to our property during one of their many migrations, the rounded brown shells of the snails could be heard crunching and snapping underneath our tires. It was like driving over a road of seashells, and the tires would be covered in a gooey mess of snail entrails. Still, the line of snails would move as a huge mass, ready to leave a swath of ecological destruction in its wake.

It got so bad we had to assign two YWAMers to go on snail patrol every night. The two conscripts would walk around the property scouting for snails, grab any snail they found, and drop them into buckets. When the buckets were full, the duo would empty the pests into a big forty-four gallon drum. The patrol would guard our humble farm, and collect thousands upon thousands of snails every night. They would end their watch a bit bleary eyed, and if they weren't careful, with an aching back!

During the day, a student was responsible to fill the snail drum with water and cap it off. Over the course of the day, the sun would heat the water until the snails were cooked and no longer a threat. The water would turn into a mucky snail juice and we'd use it as a potent fertilizer on the farmland. When God gives you snails, you make fertilizer!

Each night, we would make a dent in the snail population. But we could never eradicate them! A few local Chamorros, moved by our efforts, let us in on a secret. There was much debate at the governmental level about whether or not to introduce the flatworm into the population. The flatworm is a parasite that will eat the snail, and then use the shell as a place to reproduce. The young worms grow within a home made of the shell from the snail that the parent flatworm digests.

The legislature couldn't decide what to do, and was hesitant to import another problem to the island from Guam. Paperwork and bureaucracy held the decision to act in limbo. One day a local farmer approached us and told us where flatworms could be found on Saipan. "Go down to an area of the Island called Copper Field and look under the rocks," he advised. "You can find flatworms there because someone has already secretly brought them over from the nearby island of Guam. Nobody is supposed to know about it except for the farmers who are sharing the solution among the farming community."

He was right. We went to Copper Field and found some flatworms like he said. We released the few we found onto our land, and within six months the snails were wiped out. They had been a problem since the year we arrived—finally we had a way to deal with the blight. Once the snails were gone, there were no longer hosts to feed off of or reproduce from, so the flatworms died off too.

The island was also overrun with another imported resident: the toad. Like the snails they also migrated across the road. Standing on our property, looking down the lane, you had a sense of what Egypt might have looked like when God plagued it with frogs. The toads weren't as pesky as the snails. They did, however, make very vivid sounds, especially when run over by a vehicle!

It was impossible for cars to avoid the creatures because the roads were covered with them. It was quite a sight! A sea of the green animals swarmed and hopped, furiously trying to escape the deadly tread of car tires. The toads that couldn't move in time would be crushed under the weight of the vehicle. Each toad popped loudly as it was being flattened. A long series of Pop! Pow! Pop! sounds could easily have been confused for fireworks! Their numbers were so great that if a car drove down the road during a wet afternoon, one could have imagined oneself in a Fourth of July celebration in the United States.

Chickens

We also had chickens on our micro farm. We didn't have much money to work with, so we trekked out into the jungle and cut down what trees we needed to build a poultry house for the birds we would import. The only kind of tree available for our use was the aptly named ironwood, which is extremely tough to cut through. We felled them, and dragged them up to our property. No easy task. We had to pull the heavy trees by hand over rugged terrain to our buildings. Six-inch nails would bend when we tried to pound them into the wood with a hammer. To build our structure, we had to use a drill to make holes in the wood before we could use the nails. If the wood was green and young, then it was possible, sometimes, to get a nail into it. But when the wood dried, it was impossible.

We imported one hundred day-old chicks from California to start the chicken farm. At the time, most of the eggs on the island were imported from the USA, so they had to come thousands of miles by boat before they could be eaten. As a result, the eggs had a relatively short shelf life. We reasoned that we could introduce fresh eggs to the island, and a self-sustaining business for locals by starting egg farms.

The business really started to take off! I told Loren about our initiative and he introduced me to a businessman from California who had the largest chicken farm in the world. We told him about what we had done. He was extremely impressed with the effort we had undertaken to make the farm a reality. How we had gone into the jungles to chop down the ironwood and the way we carefully cut bamboo to create troughs to water the birds. He donated about thirty thousand dollars to us so we could expand our chicken operation. This money enabled us to build a chicken coop that could house eight thousand laying hens.

That chicken farm still exists today, after all these years, and is still in operation. We used the farm to teach Micronesians about business, finance mission efforts throughout the islands, and teach the benefits of having a good work ethic to our DTS students we were training.

Along with chickens and crop growing, we also imported six dairy goats from Los Angeles. We fed them by breaking off branches from the tangan-tangan trees that covered the island. Alan Lim, a Singaporean who was small in stature compared to us Westerners, joined our ranks as a DTS student. One day Alan went out with Matt Rawlins to gather food for the goats.

Matt is over six feet tall and a lot heavier than Alan. He could jump up and pull down the higher branches of the trees. Alan helped to hold the branch while they broke off the leafy foliage together for the goats. On one occasion, before Alan could let go of the branch they had just stripped, Matt let go. The branch had so much torque that it launched Alan's small frame into the air and left him dangling in the tree, high off the ground. On days when we needed a laugh we could easily be brought to tears in laughter by remembering the image of Alan sailing through the air, catapulted by the branch of a tangan-tangan tree. Word spread fast to the students and staff to beware the dangers of goat feeding.

We had much fun serving the Lord together, and many practical jokes were played on each other. Perhaps the only time Alan made us laugh harder than the tangan-tangan catapult incident was when he went in search of mountain raisins. He was instructed along with the other students that the raisins were very valuable, and could only be collected at certain times of the year.

A special point was made of informing students that on no account were they to eat any of them, making very sure they understood. They were given a short period of time to do the collecting. When Alan came back, he showed me the few he'd managed to find.

I looked in his container of small brown berries and said, "Alan, please tell me you didn't eat any of the mountain raisins."

He kicked at the ground and sheepishly admitted he had tried one. Hardly able to hold back my laughter I asked him what the raisins tasted like.

"Sort of like licorice," he said, with a funny look on his face.

I nearly split in two with laughter. Curious, other students gathered around. "Alan," I hooted, "those aren't raisins. They're sun-dried goat droppings!"

Chapter 26: The Stonefish

One thing you learn as you grow older is that some people take life too seriously. As we work to accomplish the goals set before us, and go about our daily lives, we can forget to have fun. All through the years, and even to this day, whenever we prepare and then lead teams to do mission work, we like to create an atmosphere where we work hard, have fun and enjoy one another.

We encouraged such important things as visits to doughnut shops or Starbucks, hot springs, playing on sport teams, swimming, board riding, volleyball, ultimate frisbee or anything else that helped produce healthy interaction. If fun and laughter in community is absent, it's usually a sign that something's wrong.

One of the fun things we loved to do in Saipan was to head to the beach on our days off. In the summer time, the island's dry season, a particular tree would flower into spectacular oranges and reds. The locals call the trees Flame Trees. All throughout the Micronesian islands it's as though every tree is alight and on fire. It was a truly spectacular sight to see. And the beaches! White sand glistened in the sun beside the clearest green-blue water you've ever seen. Coral, tropical fish, and the incoming tide drew us to the ocean.

Some days it felt like the water had been heated and the ocean was warmer than the air temperature! But our team loved to swim no matter the water temperature. It was a fabulous place to be. On one of these wonderful days off, we headed to the beach to play our favorite game: King of the Raft. We enjoyed the fun we had with each other, as we jostled for dominance of the raft. The last man standing, having thrown all other competitors into the water, was declared the King. The game was always wild, fun, and a great way to do team building.

We used a raft that was moored outside of the Royal Taga Hotel. One memorable day we found the raft stuck in the sand, waves had knocked it off its moorings and it had floated to shore. The whole apparatus was stuck in the sand, so we decided to try and get it loose. We had to wait for the tide to come in so that the waves would lift it off of the beach because it was too heavy to lift on our own. As the waves started to roll in towards the shore, a few of us gathered around the raft.

I had a good, solid grip and rapidly moved my feet under the raft so that I could lift it straight up without hurting my back. All of a sudden, a searing pain shot through my foot. The pain was agonizing, and it felt as though someone had thrust a knife into the end of my foot. I let out a loud yelp. I had stepped on a sea urchin before in Hawaii, which was painful too, but nothing close to this level of pain. I could feel sharp barbs piercing the ends of my toes. I screamed out again and let go of the raft.

I hobbled to shore and lay down in the sand, barely noticing the waves that crashed against my body. In a matter of seconds, I was battered by an onrush of pain, which rolled over me like the tide. The acute pain was intolerable, violent and like nothing I have ever experienced in my life. Matt Rawlins and another teammate helped to pick me up off the sand. They placed my arms over their shoulders so I could take weight off my foot. We got to a car as quickly as we could and they rushed me to the hospital.

The diagnosis wasn't good. After looking at my foot the doctor told me I had stepped on a dreaded stonefish, the most venomous fish in the world! This fish spends most of its time on the seabed, and is camouflaged so that it looks just like a stone. It was hiding under the raft and out of sight. When I pushed my feet further under the raft to gain a better footing I had inadvertently struck the nasty fish. Its defense mechanism is to use spiky dorsal fins to inject venom into anything it perceives as a threat. The poison immediately begins to break down any tissue it comes in contact with, leaving the hapless victim in horrendous pain.

The hospital staff didn't know what to do. There was no anti-venom serum anywhere on the island for this poison now coursing through my body. The only place it was available was on the Gold Coast of Australia, which was thousands of miles away. The nurses observed me for a while, but since they didn't know what else they could do to help, they released me. For treatment, all they could offer me were painkillers to help with the unbearable pain still coursing through my body. They sent me home with a bottle of pills.

The painkillers didn't even come close to reducing the pain I was in. I was completely beside myself in agony and desperately grabbed my foot and squeezed my ankle in an attempt to stop the pain. The pain was so intense that crazy thoughts ran through my mind like cutting off my foot! Matt brought me back to the hospital as I writhed in agony. "I can't stand this pain," I said to the doctor, desperate for any other treatment they could give. "This is intolerable!"

The doctor said there was nothing more they could do. "You're already at the maximum dosage for pain medication," he told me, not unkindly.

"What about Novocain?" I asked, "Can you shoot my foot up with that?"

The doctor scratched his head and looked at the nurse. "I hadn't thought of that," he said. He prepared a syringe and injected my foot with Novocain, and finally the nerve-splitting pain in my foot was numbed.

I went home relieved. We learned the poison that comes from a stonefish is a protein that starts to break down cells and nerves the moment it enters your body, hence the terrible pain. The quills were lodged deep into my big toe and the doctor could do nothing right away to extract them, so poison continued to spread through my foot. Soon after I arrived home, the anesthetic started to wear off, and the numbed pain was again unmasked. I quickly returned to the hospital.

The doctor injected my foot with more Novocain and then decided to cut off the top of my toe in order to get at the quills that were deeply lodged in my foot. The thick needle-like quills from the fish are not like the thorns of a thistle or a rose stem that are stiff, and can easily be extracted. These quills become jelly-like and break down once embedded in flesh.

By the time the doctor started to cut at my foot, my toe had swelled to the size of a potato. It was ghastly and shocking. My whole foot had turned black and purple and was swollen and misshapen. It was a ghoulish sight. A dark scarlet mark trailed up a vein from my foot all the way up to my groin. The doctors told me this was the poison from the stonefish.

My flesh was discolored and my foot remained swollen for many months to follow. It was eleven full months before I could put a shoe on my foot again or even see my ankle. Even though the pain could be numbed I wasn't off the hook.

Twice during the following months my body was thrown into terrible fever as it tried to fight off the poison that worked its way through my system. One of those times, during a short trip to Malaysia, the doctors there didn't know what to do and suggested I be rushed down to Singapore. Here I encountered one of the few doctors in the world who had ever successfully treated a stonefish patient. He put me on mega doses of antibiotics, in hopes of flushing the poison from my system.

A few weeks after that fateful day when we tried to move the raft, a Japanese man swimming in the ocean stepped on a stonefish as well. The venom was released in the central part of his foot. Within hours of the incident, the poison had travelled up his body and paralyzed the man. He was airlifted back to Japan for treatment, but along the way, his vital organs shut down, and he died.

The doctors said that I was lucky, that by some chance I had been poisoned at the end of my toe. Had I stepped on the fish mid-foot and been pierced with more quills and more poison, I too would probably be dead. It was a close call, and I'm convinced God spared my life. I was carried through those horrible days by the prayers of my teammates and friends.

Missionaries have to contend with many things while they are at work on the field. My near-death experience did not occur while I was out preaching the gospel, but on a day of rest as the stonefish sidelined me.

In the years to come I would experience real tragedy—the lives of friends dear to me would be stolen from them as they went about the work of serving God. There would be a few other close calls on my life, and on the lives of those that I love. All of these experiences served to teach me how dependent, at all times, we are on the grace and mercy of God.

Despite Alan's early disobedience and his affinity for mountain raisins, I later proudly handed over the leadership of the base in Saipan to him in 1983. He was a good leader who loved the Islands and the Island people. He demonstrated excellence in character, was always loyal, hard working, energetic, competent, mature, able to teach, full of vision and integrity, and always had a twinkle in his eye. As I passed one baton on to Alan, another baton was handed to me. Ross and Margaret Tooley were leaving the Philippines and heading to Kona. The Asia Pacific Leadership team had asked me if I would be willing to return and head up the ministry there.

After praying I wholeheartedly said, “Yes!” I was ready for a new phase in life and thrilled to go back to the country and people I had come to love. Returning to the Philippines was like returning home. Though I didn’t know it at the time, it would be there that I would face some of the greatest challenges in my life. The Philippines was also the place where I would fall in love and enter into a whole new phase of life: family and fatherhood.

Chapter 27: Return to Philippines

In 1983 I returned to the Philippines to direct YWAM's work there. Among the early teammates who came with me from Saipan were Jeff and Donna Romack, John Parsons, Christopher Weiman, Bill and Mona Harwood and John Saison. When we arrived at the base in Baguio City, we inherited a small team that Ross and Margaret had been leading and discipling. They were a fantastic group, and one of the first things we did as a new team was to hold a staff retreat.

We met together in a retreat facility in San Fabian, the low lands of Pangasinan, which was owned by a wealthy family from Manila with whom I'd become good friends. Together our team size was seventeen; five were from Baguio, the rest had come with me. We gathered to spend time together in fellowship, get to know each other, and build our new team.

It soon became apparent to me that it was time to move and expand the work of the mission in the country. The ministry out of Baguio City was focused primarily on the mountain tribes. Jeff and Donna Romack stayed to continue the work among mountain tribes in the north. I moved with the majority of the remaining staff to Metro Manila where we established the national office for YWAM in the political and commercial nerve center of the nation.

We had difficulty finding a place to base our work at first. Kim and Kevin Darrough were leading a School of Creative Arts, and had brought a team of eleven enthusiastic young people with them on outreach from Honolulu to Manila. The team was working with a church that was situated beside a garbage dump in Metro Manila.

At night, the team slept in a church, curled up on individual pews. There were mosquito nets all over the place like a miniature tent city inside the sanctuary because the church windows had no screens. The team struggled through each night, sleeping on hard wooden pews, fighting to keep flies and mosquitoes out of their nets. They were struggling the same way I had over ten years earlier when I first arrived in the country. It was more than evident that YWAM needed a place of our own to house staff and visiting outreach teams.

I informed Kevin and Kim's team that we should look for a place of our own to rent. Not too long afterwards, Kevin came to me and said that he and Kim had found a little house in the Nagtahan area. "It's rather small," he told me, "but at least it's a place. Perhaps we can start the work there."

I went to take a look.

"What do you think?" Kevin asked me when I returned.

"Well, it definitely is small, and humble," I said, "But we're desperate for a space of our own, so lets take it and see if we can make it work."

Soon we were renting the little house. It had two bedrooms upstairs and one bedroom downstairs. There were twelve of us who stayed there in total. Kevin and Kim stayed in one of the upstairs bedrooms, five of us single men stayed in the other upstairs bedroom, while five single girls stayed in the downstairs bedroom. The rooms were tiny, the size of a very small office. Each of us guys had an air mattress on the floor, a mosquito net to pull over ourselves, and not much else. To get in and out of the room we had to weave through countless cords strung in all directions not unlike a spider web. The cords were holding up all the mosquito nets. My briefcase was my office and I did much of my administration work on the floor.

We had a lot to be thankful for, however. With a roof over our heads and four walls to surround us, air mattresses to sleep on and netting to keep away pesky insects, we had more than most people in the area could dream of. Metro Manila was riddled with small barrios, or villages, made up of families often with no work or income, who lived in utter poverty. They squatted in dangerous places: alongside railroads, riverbanks, and under bridges, creating little neighborhoods where no one else would ever choose to live. At the time there was an estimated three million people who lived like this all throughout Metro Manila. We called these areas Squatter Zones.

Nagtahan was squatted on the edge of the Pasig River, which at some point in its distant history was probably a clean flowing river. When we arrived in the area, however, the river was a wretched swath of water, so polluted that it flowed like black sludge. The river was close to the mouth of Manila Bay; thus, it was tidal. For the most part there was enough current to keep its putrid smell from overwhelming our senses. The river, filled with garbage and refuse, sadly emptied into Manila Bay. Scientists did water samples of the Bay in the

1970's, discovering every known communicable disease festering in the Bay, largely because the Pasig River fed into it.

This barrio at Nagtahan squatter zone was jammed against the river on one side, and a major thoroughfare on the other. To draw one of the most stark contrasts imaginable, just a block or so to the south was the opulent presidential Malacanyang Palace. The government had put up trellis fencing everywhere to contain the slum, and to block it from the view of passing motorists. It was an eyesore officials preferred to forget about. The residents of the squatter zone made huts out of anything they could find: wood, tin, cardboard, plastic and rubber tires. They used rocks to hold their roofs in place. The huts were jammed into every available space; thousands of people lived in the small, wretched area.

This village stole its electricity from nearby power lines, running makeshift wiring to little huts in order to power what electronics the inhabitants had salvaged, like radios, toasters, and cassette players. It didn't matter how poor the residents were, it was amazing to see that many of them had TV antennas! The locals climbed up the power poles, trying to tap into a power source: such a dangerous endeavor. It wasn't uncommon for men to get electrocuted in their attempt! There were probably several thousand huts in Nagtahan, representing multiplied thousands of people living in squalor. Residents were lucky if they had food, electricity, clothing, or a shack to live in.

Nagtahan was YWAM's new home and neighborhood. There was so much need amidst the poverty—both in the physical sense as well as in the spiritual sense of the word. We decided to walk through the slum, going from hut to hut to meet our neighbors. One of the first things we noticed, to our shock, was that all throughout the barrio the children were badly malnourished and sick, their lives hung by a thread. In many cases, there were no adults to watch over them during the day.

"It's unbearable!" one of the girls on our team said in tears. "To think, all these kids are just slowly dying of hunger and neglect."

I agreed. "We have to do something," I said.

We asked the older children where their parents were. In the majority of cases, the mothers had to leave their kids to go to work very early in the morning. Most often, the women washed clothes by hand in the city. They worked hard, long hours each day, away from home and were paid hardly anything. They made barely enough money to buy rice for the family. Most huts had one or two children with another child on the way!

Virtually all of the husbands were without work. Alcohol abuse was very rampant among the men in Nagtahan, they would spend what little money the family had on alcohol. The majority of husbands and fathers just walked away, neglecting and abandoning their families completely. These men left their wives and kids to fend for themselves, eking out an existence in abject poverty.

It added to their hardship that many of the women were pregnant. They had to work and strain their bodies as new life grew inside of them. The women trudged through each day with the inevitable dilemma of having another mouth to feed! It was a regular sight to see a mother leave her hut, pregnant and with an infant strapped to her back. She had no food in her stomach and a twelve hour day of work ahead of her which would produce hardly enough money to feed the child she left behind all day at home, let alone her own tired and hungry pregnant body. It was an unimaginable life.

Unemployed adult neighbors would occasionally check up on the children who were left behind by working mothers, but it wasn't frequent. We witnessed many children all alone, tummies bloated with third degree malnutrition, their hair discolored because of the lack of food and nutrition, with no clothes on, staring vacantly off into space. Most were very young, maybe eighteen months old at most. What we saw broke our hearts.

One of the first neighbor children we encountered was so malnourished and in such a life threatening state that we rushed her to the hospital. We told the neighbor in the hut next door, "If this little girl doesn't get help immediately she'll die." We asked the neighbor to tell the mother when she returned from work that the missionaries down the road had taken her to get medical attention.

Before we went to the hospital we detoured into our small home nearby. We put the baby under the shower, since she had been lying in her own filth. We cleaned her, wrapped her in a towel, and then two of our staff took her to the hospital. When our staff arrived at the emergency room and asked to see a doctor, he looked the baby over, and quickly concluded that it wasn't worth treating her. "She's too far gone," he said. "The fact is that she will die within the next few hours. You'd be wasting your money."

The girls couldn't believe their ears! "We don't care how much it costs. This child's life is more valuable," one of the girls told the doctor, adamantly. "Please, do anything you can to try to save the little baby." The doctor finally agreed and reluctantly admitted the young child to the ward, where she was given an IV, some medication, and kept overnight.

At the time, the hospital's policy was that anyone who was admitted to its care had to have someone accompany the patient at all times. With no way to pay for the care, but hoping the life could nevertheless be spared, it wasn't unusual for people to drop a sick baby off at the hospital and then abandon the child. The parent's actions forced the hospital to become responsible for the child.

Therefore, the only way for a patient to be treated at a hospital was if they were accompanied, full-time, by a friend or relative. That person was also responsible to act as a runner between the doctor and the pharmacy, to purchase drugs and any medication the doctor prescribed. Hospitals rarely dispensed drugs because they had few in supply. Also, they didn't want to administer a drug that could not be paid for by impoverished patients.

After we brought that first child to the hospital, we went through the rest of the neighborhood, from hut to hut, only to find the same situation everywhere we went. Before we knew it, nearly all of our YWAM team was at the hospital, staying with different children who had been admitted. We didn't have a team left to do any of our other work! We soon made an agreement with the hospital's administrator that allowed one YWAMer to be responsible for several babies at one time. They would stay at the hospital during the infant's treatment, go back and forth between the hospital and pharmacy, and get medication for the children that the doctor had prescribed.

Once we had brought a child to the hospital during the day, we would return to Nagtahan at night and tell the mothers we had taken their children to the hospital. Then we would give them the diagnosis that the doctors had given: that their child was expected to die, or was severely malnourished, or might have brain damage. The mothers would just weep. They already knew the situation was terrible. They had watched their children's health slowly deteriorate day by day as their precious babies slipped into worse and worse health. But what could they do? They had no money or means. They did the best they could to keep their children alive, they were thankful for the help we gave.

Incredibly, all of the babies we brought to the hospital recovered fully; we didn't lose a single one! Several times when the babies were due for release we didn't have enough money to pay the bill, so we'd all be on our knees praying for money. Every time, money would miraculously arrive that very day, not once did we ever have a baby remain in hospital one day longer than necessary through lack of funds. God cared about those precious children and enabled us to keep on top of all the bills.

We happily brought them back to their mothers who joyously received their children as though they were raised from the dead. Soon the cycle would repeat itself, however. With little food to give their children, they would regress, returning to a state of health as bad or worse than the state they were in before.

We hated seeing this cycle play out. We needed to do something for the community to help the mothers and children get out of their degrading and unhealthy state. It wasn't enough to meet their immediate needs, there had to be long-term change. It wasn't long before God planted an idea that would transform the Nagtahan squatter zone along with much of the future ministry YWAM would do in Manila.

Chapter 28: Nagtahan Day Care Center and Kindergarten

Soon after we had settled into the area at Nagtahan, we found another house across the street. We rented the small space and rolled up our sleeves. We cleaned, painted, and prepared it for our exciting new project: a day care center for the children who lived in the slum. We started a soup kitchen in the building as a way to provide proper nutrition for the children attending, as well as for their mothers. We baby-sat the children during the day while their mothers went to make their meager living in the city.

The mothers came to the center early in the morning, around 6:00 AM, to drop off their children. About 6:00 PM, their mothers picked them up on their way home from work. These mothers were so tired after their day of washing clothes that they headed straight home with their babies to sleep before waking early to repeat the same routine.

We started with about twelve to fifteen babies. It was all our small facility could handle; of course, the need was greater. The soup kitchen, however, we opened to all the mothers in the barrio without enough food. Many of the families could only afford rice, and only occasionally a little bit of fish. Each day we would give high protein meals to the mothers as well as to their malnourished children.

Once we started feeding the children in our care three meals a day, with snacks and treats in between, and bathed and clothed them regularly, we noticed massive changes in them right away. We gave them heaps of love and cuddles, and it was like watering small withering plants that had been abandoned in a garden—they seemed to spring to life!

Most of the children, who, by that time in their development, should have been walking, could not yet walk or stand up on their own. We had nurses on our team who worked with the children to do physical therapy by moving their muscles and pumping their little arms and legs. Physiologically, every one of them had major developmental setbacks. So our nurses and staff gave them excellent care. We gave the children whatever therapy they needed, together with daily baths, good food and clothes. Most important of all was the unconditional love our nurses gave them. The children all responded fantastically.

As we started to meet the basic needs of kids by getting them on track to a normal life, we soon perceived that the children would never have a chance to go to school, even though they were now healthy and strong. When they were old enough for school, they would have nowhere to go. Thus, we started a kindergarten for the children of Nagtahan.

The families of the children had no money, so none could afford to pay the costs of the school. The building we were using was in the squatter zone itself, in the midst of extreme poverty; but it contrasted dramatically with the environment where it was set. We painted the space beautifully, and made some wonderful educational toys. It was bright, cheery, clean, and pleasant to be inside.

The kids' eyes lit up with wonder the first time they walked into the school. Its simple and sturdy tables, its shelves with books and toys in bins they could play with, all appeared so fantastic. It wasn't long before word got out about the place. We developed a very good reputation in the area. Soon rich families were coming to us to ask if they could enroll their children into our kindergarten, even though it was in a slum!

We decided we would take five children from wealthy families and charge them an arm and a leg to attend. These parents were happy to pay because they were convinced that we had the best kindergarten in Manila. The money from those wealthy families helped to subsidize the costs of the fifteen children from the Nagtahan slum.

Church Under the Bridge

The people of Nagtahan were overwhelmed by our kindness. Why would complete strangers from wealthy countries come and give their time, energy, and money to help them in their need? Why would we bring their children to the hospital for the medical attention they required and pay the bill? Why would we do this when their own people or government would not?

It was the love of God that compelled us. For so many people, the love of God seems a thing that is very far off. How could God even exist if there is so much poverty and war? How can God exist if there is so much human need? And if He does exist, how could He be a loving God given the hurt and enormity of human suffering? I experienced early in my life that the love of God is real and not far off, and it has been the great privilege of my life to share this truth with people all over the world.

The love of God came freely to us, and it reaches into the hollowness of our broken hearts. Jesus came to earth and put this love on display. He tenderly touched the sick and healed them. He went to crowds who were hungry for food and fed them. Then He turned to those who followed Him and said, "Go into all the world and do the same." This compelling love—to minister to the hurting, the broken, and the sick—is at the very heart of missions. It is why I have traveled around the world. It is why I have preached and taught, gone into hill tribes and slums, befriended world class athletes and government officials. This Heavenly love is the reason I've laughed, cried, and prayed.

Once our work in Nagtahan was established, and the people of the impoverished barrio started to see their lives, and the lives of their children, transformed, they began to ask us "Why?" Hope was sparked in hearts that had long since abandoned it. There was something about our lives that attracted them. They wanted to understand what it was.

"How about we have some nights of public meetings where we can share with you about the great love of God?" I asked. "Would you like that?"

They were eager to hear.

One day after looking for a suitable place, the Lord gave Kevin Darrough a brain wave. What about using the space underneath the Nagtahan bridge beside the squatter zone? It was close to where the residents lived, covered should it rain, and it was large enough to fit the whole community should they all attend. It turned out to be a fantastic idea.

Malacanang Palace, the official residence of the President of the Philippines, was a ten-minute walk from Nagtahan. It was always such a shock to the system to hit the street, walk for a few minutes away from the piled garbage, stray dogs, and malnourished residents to abruptly arrive at the entrance of the palace. In an instant the surroundings were shiny cars, fashionably dressed people, and homes of the rich.

Kevin made a trip to the palace, somehow got a meeting with an official, and asked for permission to use the space under the bridge. The palace used the space to store their security vehicles. To our utter amazement, they agreed to move them. The government's "garage" would be converted, temporarily, to a place where we would host open-air meetings to share the good news of God's love to the people of Nagtahan.

Initially, it was the mothers and children of Nagtahan—those who came to our soup kitchen, Day Care, and kindergarten—who attended our revival meetings. A few curious teenagers also attended. They were all very open to the gospel. Virtually all of those mothers believed the story we told of God's love, and accepted Christ into their hearts.

We continued meeting there until we could find a more permanent location. This gathering of believers came to be known as the Church Under the Bridge. Eventually we were able to rent additional space right beside the day care center. One of our team members became the church pastor, and the name changed from the Church Under the Bridge to the Nagtahan Church. Much discipling took place in those early days; the church was off to a great start. As it grew and matured, we were looking to install a particular brother as an Elder. He was the barrio captain, quick to learn and eager to grow.

As we went through the scriptures with him, teaching the qualifications required to function as a church Elder, he shocked us by telling us he had three wives! No one had any idea he was a polygamist. As we continued to disciple our new converts, we ran into all manner of issues that needed Biblical answers. You can be sure we were kept on our toes! Constantly, we were challenged and stretched. But this was God's work, and these were His people. God was well able to grow what He had planted. He gave us the grace and tenacity to press on.

Some years later I returned to visit that little church, we had a great reunion! The Nagtahan Church continues to this day.

Chapter 29: Malakas Street and Smokey Mountain

After being in our small house in Nagtahan for about six months, the other YWAMers and I learned of a new place in Metro Manila available for rent. Compared to the place we were renting in the slum, it was an absolute mansion! The house was situated in a quiet residential area in Quezon City on Malakas Street. It was a gated property with a guardhouse. A U-shaped driveway curved under a sweeping canopy, which welcomed arriving guests to the impressive home, and invited them to climb the beautifully polished marble stairs.

The home was built for the Minister of Foreign Affairs of the Philippines. It was designed to be a place of welcome and hospitality for international diplomats. The mansion seemed a fitting place for us missionaries, who had come to learn, like the apostle Paul, to cope with little or much—in slums or in a mansion. We too were like ambassadors in the nation, only of another kingdom. Our responsibility was to display God's love and show hospitality to the indigenous people with whom we wanted fellowship. Only God could have opened the door and worked out the necessary circumstances for us to acquire such a property.

Despite its fantastic design and impressive history, the property was little desired by Filipinos. To them, it was jinxed and cursed. The previous owners of the home had been burgled and one of the guards watching the residence had been murdered during the ordeal. At the very least, it was considered bad luck to live there, but more readily it was seen as taboo. Nobody wanted to live in a place they perceived to be haunted—hence the cheap rent!

We refused to come under fear of any curse, and we claimed the place for God's work. The family who owned the property was happy to make some money off their investment. It was much less than they made prior to the break in and murder, but they were happy to have it occupied by tenants once more whatever the profit loss. We had a new headquarters for our growing work in the Philippines.

When we were settled and organized at Malakas Street, one of our young missionary couples came to us. George and Tracy Rankin worked in Tondo, the oldest city in the greater Metro Manila area. They had started a base in the city, and came to talk about their progress with the new plant. Both were courageous pioneers. Tondo was riddled with gangsters and thugs. Police refused to go into the area after sunset, and would not enter before sunrise. It was filled with crime. Tondo was not welcoming to the faint of heart, and this young couple called it their new home!

I could see the passionate urgency in their eyes when they arrived. It's a look I've seen in so many young missionaries over the years. "Kel, did you know there's an island right off Tondo made entirely out of garbage?" George asked.

"No, I didn't," I replied, surprised.

"There are many thousands of families living on it," Tracy said. "The island is about seven hundred feet high of smoking garbage, the result of over two million tons of waste heaped right there into Manila Bay."

George explained the trash island was totally unfit for human dwellings, and yet people not only lived there, he said, but made their living from it as well!

"You should see it." Tracy spoke up again, "It's terrible! Nobody should have to live like that".

"We've been all over the island," George said, "and now to make things worse, the rainy season is here and the place has turned into a perpetual bog of sludge. Paths are no longer visible and the only safe way to protect your feet is to wear gum boots."

"It's worse than bad," Tracy frowned. "The way of life for the people on the island is inexplicably horrible, Kel."

"The smell says it all," George said. "You won't believe it. It's an ever present stench of rotting garbage."

The couple continued to describe Smokey Mountain. It sounded like something out of a movie. There was no running water, electricity or sewage treatment on the island. People on the mound of trash lived off of the garbage, collecting and scavenging anything that could be of use to them. Like the people of Nagtahan, they made huts and homes out of whatever discarded pieces of junk they found useful.

Kids walked all over the island with steel rods. They thrust these into the ground until they hit something hard. When they did, the kids pulled away the garbage with their hands until they found the hard object they had tapped with the metal rod. This is how they discovered bits of tin, glass, rubber, metal, and hard plastic. At the end of the day, they gathered up their findings in piles and carried them back to their homes in bags.

I agreed with George and Tracy that YWAM should help the people on the island in some way. We decided to check it out to see what we might do.

At this time, another very gifted and capable couple, Dave and Mary Anderson, joined our team. They relished the opportunity to minister to the poor, and took on the huge challenge that Smokey Mountain represented. One of the first things they discovered was that all the residents on the island had rampant parasitical infestations. One of the young YWAMers working with the Andersons on Smokey Mountain showed me a photograph. It made my stomach churn. I saw a little boy, squatting beside a pile of garbage to relieve himself, with worms hanging out of his anus. An unforgettably hideous sight.

The whole island was awful. It wasn't an uncommon sight, early in the morning, to discover a dead body. Often we saw a body left in a heap, like decomposing trash, discarded under the cover of the previous night's darkness. A victim murdered for crossing the wrong person. All dignity had been stolen from the pinnacle of God's creation, and was now reduced to worthless garbage! Death and disease were the norm. The needs of Smokey Mountain were enormous.

Dave and Mary kicked off our efforts with a de-worming drive. They set to work to relieve all the residents of the terrible parasites. The ministry on Smokey Mountain, like the rest of the work in the Philippines, began with young people noticing needs in a community, and then responding to them. Our efforts to de-worm the residents on Smokey Mountain led to more and more ministry opportunities. Our team of health care workers, under the capable leadership of the Andersons, ended up vaccinating the entire island on three different occasions for various problems.

Inevitably, as we met people's physical needs, they were so warmed and blessed by this unexpected kindness that they asked us, "Why are you doing all of this?" We then shared the truth about God with people who were touched by His love through our hands.

Additionally, we used our feet. Our teams walked all over Smokey Mountain to assess and meet needs. Next, we sent out Primary Health Care teams. Each person on the team wore a homemade medical kit, its appearance not unlike that of a popcorn salesman at a baseball game. It was tiered with compartments for bandages, pills, antiseptic ointments, syringes and various other prescribed medicines.

Instead of having people come to a clinic, which many were reluctant to do—worried about having to pay money—our teams chose to go to them. It was also a way to ensure that the people who needed the medicine got it. This was particularly true of patients who had tuberculosis. If we dispensed the medicine, but didn't administer it, often residents of the Mountain would sell it. It was a quick way to make a few dollars at the cost of their own health.

A number of routes were developed, and the team made regular rounds. Each member of the team had a specific area they were responsible for, and would go from hut to hut taking care of the basic health needs they were confronted with. Each team made sure each patient took his prescribed dose of medicine. This was a good way to get to know the people on the dump, to build friendships with them. Our workers learned so much about the people by interacting with them and taking time to listen.

I remember one conversation I had with a father on Smokey Mountain. I asked him about his life, what he thought of living in such a horrible place, and he said, "This is our destiny. We'll always be here and our children after us. It must be God's will."

Those words appalled me. Created to live in garbage? My destiny is to live in a dump? I was incensed to see how the devil had so shrewdly convinced the people on Smokey Mountain they were no better than garbage dwellers. What a lie! They were made in the image of God. What an honor it was to tell him otherwise. "Friend," I said as my heart beat like a kettledrum in my chest, "you were created for so much more than this. This dump is not your destiny. It is your destiny to be a child of God! You don't belong here, and none of these other people do either. God has something vastly better for you and your children!"

Chapter 30: Coffins for Sale

The work in Smokey Mountain flourished under the leadership of Dave and Mary Anderson. They arranged for local doctors and YWAM nurses to come into the area to start medical clinics. They also began micro industries to help the people earn money to turn their lives around.

In Tondo, death was a daily event. Life was brutal and harsh; death seemed to touch everything. Illness and disease in the area claimed countless lives. Because families were so poor, the price of proper burial would cost the grieving families dearly. No one had the money necessary to pay for funeral arrangements. Commonly, grieving family members begged aunts, uncles, neighbors—anyone they knew—to give them money to bury their dead.

Families were heaped in shame by not burying their loved ones properly. But if they were successful in borrowing what money was needed for the funeral arrangements, especially if it came from 'loan sharks', they were thrust into a debt that they spent the rest of their lives paying off. It dawned on Dave: we could make coffins much cheaper than we could buy them! He went to a lumberyard to price some wood.

Dave purchased saws, hammers, chisels, screwdrivers, planes, sandpaper, varnish, and some appropriate brass fittings and went to work making coffins. He got some local women to line the inside with some satin fabric. When they had finished, they looked as good as the coffins for purchase! They had made the same coffin for a fraction of the price of the coffins sold by the undertakers. YWAM also performed the burial ceremony for the grieving family. When the family realized they didn't have to go into any debt to bury their dead, they wept with gratitude, they were so overwhelmed they felt they didn't have enough thanks.

This experiment birthed a small industry in the Smokey Mountain area. Our team taught a group of Smokey Mountain men how to make caskets. The team gave the men the tools Dave had bought and they built a little shed right there on the garbage dump. It was used as a workshop. Our team gave the men some money to purchase supplies. They took those who wanted to learn the skill of coffin making down to the lumberyard to show them where to buy the necessary wood. Soon there was employment for men—word spread throughout the whole area quickly.

The coffins were very high quality, and so much cheaper than could be found anywhere else. Rich people from the city started coming out to Smokey Mountain to purchase the coffins because of the quality and affordability. People who lost loved ones came to Smokey Mountain to purchase a nice casket for a fraction of the price. A small casket-making business was born. It became a very successful enterprise. Dignity and self worth returned to the workers as their hard work was rewarded with good income.

From the Ashes, a Bright Future

There were over five hundred children on Smokey Mountain not going to school. These youngsters spent their days poking through the garbage, helping their parents find bits of glass, rubber, or whatever might have a market value. The kids brought back the junk they found, and piled it outside their hut. When they had collected enough, the parents brought the valuable garbage to Chinese merchants. These merchants purchased the useful garbage for a small fee and recycled what they bought.

The families made just enough to survive. They never had enough money to send their kids to school. Their children would be destined to live in the dump, with no skills or opportunities other than sifting through garbage.

To attend school, all the children needed were uniforms. The families of course did not have the money to purchase them. While pondering this dilemma, Dave and Mary's team came up with a possible solution. "Why don't we send out a newsletter and see if we can raise enough money to buy some Singer sewing machines?" Since there was no electricity on the island we needed the manual treadle type. "If we get some machines, we can teach the mothers how to sew." And sew they did. Through that one newsletter, enough money came in from generous friends, families and supporters in churches to purchase twelve brand new Singer treadle sewing machines.

Once we had the machines, the team took some of the mothers down to the market to buy the fabric they needed for the uniforms. The mothers did the bartering, got the lowest price possible, and we brought back the material to the dump. The YWAMers who knew how to sew taught some of the mothers how to make a pattern, and how to measure the cloth appropriately, while other women were taught how to use the machines.

Pretty soon we had a whole assembly line going. Our team built another shack right there on the dump where the women could work making school uniforms. All of a sudden, Smokey Mountain had another booming industry. Before the next school year was due to start, we had many mothers from the dump involved in the sewing endeavor. It was a project for the whole island. All the five hundred kids now had uniforms, but none of them had bags to carry their supplies. So our team came up with a design and pattern for school bags and taught the mothers how to make five hundred little school bags for all the children. We supplied them all with paper and pencils.

What a wonderful sight it was to see on Smokey Mountain that first day of school! Early in the morning there were kids coming from all directions, descending down the murky trails of smoldering garbage, all smartly dressed in their new school uniforms. Every child had a bag slung over his or her shoulder, with a proud bounce in each step—ready to learn, walking into a new and promising destiny. There wasn't a dry eye in the village.

Parents stood outside their huts and watched their children march off to school, proudly waving. They watched their kids whom they never thought could have a chance to learn, walk to their first day of school. Our team stood there just imagining all of the angels in Heaven rejoicing as they watched these kids. Some were as old as seventeen, who had never studied a day in their lives. As our team watched the children march off to school, heads held high, they were watching a whole community transform before their very eyes.

I asked one of the parents what the difference was on Smokey Mountain after YWAM came eighteen months earlier and befriended the people there. The father thought for a moment. He looked around him where we stood outside his hut, garbage stacked and strewn, forming an unnatural landscape. Wafts of smoke trailed thin grey lines to Heaven, and he toed the ground, nudging a mound of garbage near his foot. "Before you came, we all had darkness in our eyes," he said. After a moment, he looked up at me and smiled. "Now we have light in our eyes."

Chapter 31: Not Without Cost

Just as I was seeing great advancement in God's kingdom, 1985 was also going to be a year of great pain. I was thirty-five. As the director of YWAM in the Philippines, I could see the work was being firmly established. We had workers from around the world serving with enthusiasm and we were raising up local believers to lead and do the work of the ministry.

God was moving us into exciting new territory, establishing mercy ministries to the poor, both in Nagtahan, and on Smokey Mountain. Apart from this work, by 1985 we had established a School of Evangelism, a School of Primary Health Care, as well as a School of Counseling. We regularly ran Discipleship Training Schools in Baguio City, Antipolo, and in Davao. God had helped us to launch many different schools in order to train indigenous people to be an effective missions force in their own nation. We pioneered a base of operation in Cebu, another in Northern Mindanao, and a base in Suragao as well. By the end of 1989, we had established ten bases in Metro Manila.

We confronted many forms of evil while establishing our work. Something horrendous we encountered happening throughout the country was the forced prostitution of many young girls in their early teens. We were shocked to learn that smartly dressed businesswomen from the city headed out to visit remote farming villages to offer poor farmers the opportunity of good employment for their young daughters in the big city. These businesswomen shrewdly deceived scores of parents, promising poor families that their daughters would be able to send back money each month to help out the rest of the family. These women promised to take good care of their young daughters.

Once they arrived in the big city the daughters were forced to work as prostitutes. In the city the girls only knew the women who had deceived them, and since they were far away from everything familiar, the girls found themselves trapped. They hated their lives and were filled with shame. Even if they wanted to escape, they couldn't, since they were watched closely and controlled, never allowed out by themselves.

One of our married couples, Nick and Mary Schreifels pioneered a half-way house in Olongapo, successfully helping scores of girls get out of prostitution. They trained the girls in various job skills, enabling them to gain meaningful employment. They took care of the girls' medical needs and led them to faith in Christ. The rescued girls shed many tears as they began their journey toward mental, physical, and emotional recovery.

As the girls opened their hearts to God, they quickly learned that He cares deeply for the broken hearted and down trodden. They were amazed to discover He loves and grieves over the young and the weak, those taken advantage of by the wicked of this world. What joy to see these girls set free from condemnation, and walking in newness of life as Jesus offered them salvation. Christ marvelously returned to them their self worth and dignity!

As YWAM expanded, all the new established works made for a vibrant and fulfilling time. However, it wasn't devoid of hardship. Our whole team was rocked by tragic news on one fateful day, January 26, 1985. We had already moved the national office to Manila, while the Romacks remained with a team in Baguio, working in the city and among the hill tribes of the north.

Mike and Janice Shelling had returned to the Philippines after completing further training at Bethany Bible School in the States. A small team from Mike's church in New Zealand was visiting our Baguio base, helping us with the work we were doing among the hill tribes. Although the visiting team expected the work on their short-term trip to be challenging, nothing could prepare the team, or myself, for the awful events that were to unfold.

Mike was overseeing the visiting team. The outreach had been a success and the team was preparing to head back to New Zealand. Mike had arranged to send some personal mail back with the team, but he didn't show up for the morning meeting. He was always a punctual man, so a member of the team was sent to walk over to his house—it was only a hundred yards or so from our base. It was out of character for Mike to be late, and the team was getting anxious about missing their bus back to Manila.

Around 10:00 AM the team member arrived at Mike's front door. He encountered an unthinkable scene of horror. The front door to the Shelling's home was ajar and dried blood covered the entryway. The man could hear a baby crying inside. He pushed open the door and found Mike and his wife

Janice lying of the floor, covered in blood. They were dead. Their three year old daughter was sitting on the floor between them, alive but terrified. Their four month old infant son was alone in the bedroom, crying. Mike and Janice had both been murdered in the middle of the night, brutally stabbed multiple times.

My dear friends! Their precious lives! Mike and I had been students together in New Zealand, we traveled together, prayed together, and worked together. We were best friends. I had encouraged Janice to come to the Philippines where she met Mike. It seemed as though a bomb went off in my insides. My two friends and trusted partners whom I had relied on, and had spent time dreaming with about future ministry, were gone. What horrific shadow their cruel deaths cast over us all. Our whole ministry entered into the darkness of grief. The brutality of their deaths was overwhelming.

I was in the house with the police officers as they surveyed the crime scene. There were clear footprints in the blood that covered the floor. When I asked the police to take photographs of the footprints and samples of the blood to use as DNA evidence, the policeman in charge angrily asked, “Are you trying to tell me how to do my job?” I backed off. I only wanted to ensure the proper evidence was gathered so we could secure justice for these dear friends. Sadly, the police never acquired any evidence conclusively proving who committed the crime.

The police speculated that the murderer was most likely invited into the home, as there was no evidence of a struggle or break in. The police concluded that the murderer attacked Mike after being allowed in the house. At some point, Janice, hearing the struggle in the living room, left their baby son she was watching in the bedroom to check on Mike. Then she was violently attacked, leaving their two young children without father or mother.

The police charged and convicted a man for the murders, but only God knows if he was the man who committed the crime. One of the missionaries working in a prison ministry in Manila told me that a large percentage of prisoners incarcerated in the Philippines at that time were jailed on trumped up charges. When the police needed a criminal, they’d pull a drug dealer or addict off the streets and indict them. The crimes that landed many criminals in jail were for crimes they themselves had not committed. Although they would be quick to agree they deserved to be in prison because of other criminal activity they were involved in. In those days, lazy, corrupt, or plain inept police officials forewent investigations in order to save face in the eyes of the public by putting petty drug users in prison.

There was much mystery around the murders of Mike and Janice. There was no reason for their deaths that we could discern. Drug-related crimes most often involved break-and-enters, as addicts entered homes to steal things to pawn off for drug money. There were all kinds of valuables in the house—confusingly, none of which had been stolen. The man who was charged and put in prison for their murders later mysteriously escaped! He was never caught.

A few months after the Shelling murders in Baguio, we lost another missionary in the city of Davao. This city, in the southern Island of Mindanao, became the scene of another detestable murder. On Thursday, September 25, Randy Adams was stabbed to death, alongside a Filipino maid, who had been working for their family. She was 33. Two masked men broke into their home. Randy had seen them coming towards the house and quickly sent her three girls upstairs. She told her daughters to lock themselves in their bedroom. Her quick thinking saved her girls. Randy's husband, Bill, was not home at the time. He was left a widower with three children. What outlandish violence! It was an unimaginable, heartbreaking loss.

It was a dark and painful time. Those violent events were very sobering. I have always believed—and still do believe—that being in the will of God is the safest place you can be in at any time. However, being in the will of God doesn't mean suffering will be absent. Injury and death exist, even in missions. One of the hazards of the mission field is that there are bad people in every country, no matter where you find yourself ministering. Missionaries are not exempt from terrible suffering. Through these horrendous circumstances, I had to look this awful truth in the face.

Although the people who killed Mike, Janice, Randy, and her Filipino maid did not necessarily hate their message, I consider these missionaries to be martyrs for their faith. They were in the Philippines to preach the gospel, and had they not responded in obedience to God's call on their lives to work as missionaries in the Philippines, they would likely still be alive today.

If anything, their deaths made us tighter as a community. Oh, how it required us to depend on the mercy and comfort of God! More then ever, in memory of their lives, our determination that the gospel be preached was lit aflame with a passionate fire. Mike and Janice were deeply committed to minister to the tribes in the north, as Randy was to her work in Davao. To honor their legacies, we strengthened our resolve to reach those who had yet to be reached for Christ.

Chapter 32: Pascual

"Does God really forgive sin?" the man asked in the glowing light of the small church. The burning wick of the candle reflected in his eyes. The flickering flame made his dark brown eyes seem alive, although Paulina could clearly see the man was like the walking dead. Paulina and Jacinto were a part of a team of missionaries ministering up among the hill tribes.

"God forgives," she assured him.

"Even the worst of sins?" the man asked.

"Yes, even the worst of sins."

The team had gathered for a night meeting in the small village church. "I saw the light when I was walking by. I saw the lantern hanging in the window," the young man said, his voice barely a whisper. "I heard your voices. I thought maybe I was hearing angels."

"What's your name?" Paulina asked.

"I'm Pascual. I've lived a bad life. I deserve to die. At least, I don't deserve to live."

Paulina invited Pascual into the church and introduced him to the team. As they sat down to talk, Pascual started to share his life's story. "For many years I've been in a gang. I drink a lot. I live on the edge. I've even cheated death a number of times." Pascual laughed at this, describing how he had been shot twice. One bullet was never removed from his leg. It was the result of an unsuccessful robbery. He and his gang had broken into a home. When the owner had heard the young men in the house, he chased them with a gun. Pascual escaped but a bullet ripped through his calf, lodging itself in his foot as he ran away that night.

He rolled up his pant leg. "See? Here's the scar."

"You sure have lived dangerously," Paulina said, sensing the pent up guilt he had carried on his own for years.

"You have no idea. I play this game with my friends on weekends. We take a shot of gin, pick up a gun holding one bullet, spin the chamber, point it at our chest, and pull the trigger." On one such evening of drink and dare, Pascual pulled the trigger and the gun fired. The bullet entered his abdomen. Somehow he survived.

"Why do you think you don't deserve to live, Pascual?" Paulina finally asked. A dark cloud of guilt seemed to fill the room; it hung over the young man's whole countenance.

"Why should you care?"

"Well we've really grown to love the people up here in the mountains very much, we care about them a lot, and enjoy their fellowship. If it's alright with you, we'd really like to be your friend too!"

Pascual thought about this. He looked at each face in the group—kind, willing to listen and wait. "It sounds too good to be true," he said.

Paulina responded, "It is too good. But it's also true."

After the meeting, Pascual went on a walk with Jacinto, who asked him if he wanted to take the leap of faith and receive forgiveness through faith in Christ.

"Forgiveness," Pascual muttered. "What about the worst sin, Jacinto? Will God forgive that one?"

"God will forgive the worst sin," Jacinto assured him.

"There's something I have to tell you first," Pascual said. He took a deep breath and started to share a story he rarely, if ever, shared with anybody. "My younger brother and I were always good hunters. We'd go out into the jungle to hunt wild pigs, and we'd usually go at night. I always held the gun while my brother ran ahead of me, up the trail, toward a spot where we knew the pigs liked to rest.

"One night my brother ran off ahead. I heard a noise and lifted the rifle to take aim, waiting for the pig to cross my path. I had a flashlight strapped to the barrel of the weapon, so that I could see what I was aiming at in the dark. And then it happened. I saw two dark eyes shining back at me in the light and pulled the trigger."

Haunted, Pascual spoke, his voice barely audible. "I heard a thump as the pig hit the ground, I ran to gather my prey. Only, I didn't shoot a pig. I shot my own brother between the eyes. At the funeral, my mother was totally overwhelmed with grief from the loss of her youngest son. There was no hiding the fact that he was her favorite son. That's why when I first met the team I thought I didn't deserve to live. I'm a killer. I hope some day to die the way I deserve, like my brother."

Unable to hide his desperation, turning to Jacinto he asked, "What about the worst, worst sin?"

"Yes, Pascual, even the worst, worst sin, God will forgive."

Pascual took the leap of faith that night after walking back to his home with Jacinto. He prayed for forgiveness and accepted Jesus into his heart. As the days passed, Pascual came daily to the YWAM base. He was eager to be discipled. The team discovered he was worried about what to do now that his whole life had been transformed. "How am I going to leave my gang?" The question was agonizing. It wouldn't be an easy feat.

"Why don't you join the Discipleship Training School?" Jacinto encouraged. "It begins right away in Baguio City."

"How will a school help me?" Pascual asked.

"It's a great way to learn how to develop a godly character and habits, how to study the Bible and hear God's voice. Also, it would allow you to make a whole new group of friends with the same new values."

Pascual didn't need any more convincing. He thrived in the school and never looked back. His old life was just that: old. He had been made new. Soon, he was working among mountain tribes, sharing the message of hope that had transformed his own life.

Healed and transformed by God's grace, Pascual was a strong influence. He led many people to Christ. He was a living example of the power God has to transform life. Pascual began to minister to tribes he didn't belong to, which he would have considered absurd in his former life.

Unthinkable Tragedy

One weekend, Pascual was on his way to the city of Anabel to work with Santos, a leader in the local church. Santos was from a different tribe. The night Pascual left to establish this new work, he said goodbye to the team in Baguio, and was sent off in prayer. There was great faith and expectation amongst the YWAMers at his send off for many upcoming years of effective ministry.

You can imagine the shock the team experienced the next day when they found themselves at the hospital morgue, looking at Pascual's body. It was ashen and cold, laying lifeless on a gurney. Alerted that he had been injured, members of the team took a vehicle to the hospital in Bontoc where Pascual had been rushed. They had received a note from a villager informing them that Pascual had been hurt. Expecting a broken bone, or some minor injury, they instead learned about the tragic events of the night before.

Pascual had arrived in Anabel. He was playing with the children while Santos' wife, Fermina, prepared supper and while Santos cleaned up after an unsuccessful evening of hunting. Santos changed his clothes upstairs and emptied his gun of bullets. He had put five in the gun and fired at nothing because it had rained so hard. His prey was deep in the jungle, taking cover from the downpour. Once he had removed the bullets, Santos pointed the gun at the floor and pulled the trigger. The gun went off! There was a bullet in the chamber that Santos didn't know about!

That bullet claimed Pascual's life. Fermina heard the gun go off, and saw Pascual jerk and fall to the floor. She thought he was joking around with her son; but when Santos ran downstairs to see if everyone was okay, they saw the blood. The bullet hit Pascual in the back. He was dead within minutes. Pascual died just like his younger brother.

As the team came to terms with Pascual's death, the implications of how their friend died started to sink in to the community. When a person from one tribe is killed by a member of another tribe in that area of the Philippines, the death can lead to an all out tribal war. Fear started to spread among the family and friends of Santos in Anabel. They feared tribesmen from Balili would soon revenge the death of one of their sons. War might not break out, but a revenge killing was definitely expected.

What was Santos to do? Afraid for his family, he went to the police and asked to be put in jail as protection. The YWAM workers in the region had no idea what to do; so, they did the only thing they could. They prayed for God to intervene with grace to spare the villages from more unnecessary bloodshed.

God answered their prayers through a man called Uncle Dagas. He was a highly respected man throughout the mountain province. He was instrumental in brokering peace between the two tribes. Uncle Dagas had become a

Christian on the same day that Pascual had, two years earlier. Formerly a pagan priest in the village, people came to him for advice and spiritual guidance. He knew all the proper pagan prayers and rituals. Since his conversion, Uncle Dagas had shown many people a whole new way of life, free from fear and superstition.

Soon after the tragic death, Uncle Dagas went to visit Pascual's mourning family. "Pascual would not want you to seek revenge for his life, but to forgive Santos instead." He spent time with the family, convincing them there was a non-violent solution that could bring greater peace.

Together with another respected elder, Uncle Dagas wrote out a peace pact. Pascual's family agreed not to seek Santos' life.

A peace pact like that was unheard of in the whole mountain province! Culturally, revenge was expected; in fact, revenge was the norm. The whole community of believers held its breath when a Jeep of people from Anabel arrived at the funeral to pay their respects. Pascual's family, upon the arrival of these neighboring tribe's people, offered food to eat and water to drink. This one gracious act sealed the peace. According to the tribal system Pascual's family belonged to, offering food to a visitor demonstrated that they were taking responsibility for the safety of their visitors during their visit to the village.

As a sign of their faith in the peace agreement, Santos' family pledged to give one of their own sons to Pascual's extended family as assurance that they would uphold the pact of no vengeance. So one of Santos' sons went and lived with Pascual's family. Did God not act similarly when, deserving punishment for our sin, He instead sent His son, Jesus, to die in our place? Calvary was God's peace pact.

After the funeral, Pascual's mother sought out one of the YWAM workers. She was clearly perplexed. "When my youngest son died, we went through all of our rituals—the burial, the rites—but we have never felt peace. I was angry with Pascual. I was holding unforgiveness. We did things so differently from our traditions when we buried Pascual. This time, we feel an overwhelming peace. How can this be?" The tragic death of Pascual opened up a way for the whole family to hear the Good News, for a whole village to see God's work of redemption!

Years later, Paulina, the woman who helped Pascual on his journey with Christ, wrote the following words in a letter to a friend as she faced the grief and confusion of the tragic death of the man who had become her good friend:

God is faithful and trustworthy, and sometimes we don't see the big picture. But one day, He will show us and explain to us things we don't understand. It is all about Him; I pray that His kingdom will come and His will be done on earth as it is in Heaven, and that we will follow Him faithfully to the end.

Chapter 33: A Time to Love

"It's nice to meet you Kel, I'm Kristyn." The words burned deeper into my heart every time I heard the soft voice of the beautiful woman I had met days earlier. You read about it in books, and see it happen in movies—so much so sometimes that you wonder if it's really true. For me it was. The moment I set eyes on Kristyn, my heart started to beat faster. It hasn't stopped quickening its pace each time I've heard her voice since. "She's the girl I'd like to marry," I told the Lord after I saw her. It was love at first sight.

I had been praying for the woman I would marry for years, trusting God would bring her into my path at the right moment. And it seemed the moment had finally arrived. In a way, even though I was young and unmarried, I held the role of the father for the ministry. It was time for a partner to join me in sharing the joy and challenge of life and YWAM. After all my years of waiting, this young, beautiful American girl, with blond hair and a glowing smile, appeared. She seemingly changed my whole world in a moment.

"I'm Kristyn," she said. It was the happiest introduction of my life. "I'm a DTS student."

My heart sank even as it soared. YWAM follows a clear and very good principle: during their six-month Discipleship Training School, students must refrain from dating. It's a time to focus on relationship with God.

"What am I to do, God?" I prayed. As national leader, I not only supported YWAM's rules, I also believed in the values behind them. "I can't even let her know I'm interested in her," I lamented. I didn't want to do anything to interfere with her seeking God. "This is a miserable position to be in."

As I prayed about the possibility of pursuing a relationship with Kristyn, God encouraged me that it wasn't wrong. The timing just wasn't perfect. I decided I would do what I could to connect with her and wait for the right moment, when a door of opportunity would open for me to let her know I was interested in her. Kristyn joined the work in Nagtahan for her outreach. She worked in a medical clinic for people who were living in the squatter zone. It was nice to have her nearer to me, but she needed to finish her outreach phase of her DTS before I could let her know I had feelings for her.

One day, when I was visiting her team, I learned from her leader that Kristyn was soon to leave the Philippines for more missions training in Honolulu. Motivated more than ever to find a way to show her my romantic inclinations, I sought her out.

"I hear you're leaving, Kristyn" I said, trying to appear nonchalant.

"Yes," she responded. "I'm going to Hawaii to do a School of Evangelism."

"I'm happy for you, but at the same time sorry to hear it," I admitted. "Your leaders speak very highly of you. But to be sure, the SOE is a great school."

"There's only one problem," she said.

"What's that?" I asked, genuinely concerned.

"My passport expires before I'm scheduled to leave. I just realized it and don't know what to do. I don't want to leave early and miss the rest of outreach."

"Well," I said, seeing a window of opportunity open before me, "if you want, I can drive you to the American embassy and we can make sure you get your passport extended."

"That'd be amazing, Kel!"

A few days later we drove downtown to do the paperwork. While we waited for the papers to be processed, I asked her if she'd ever walked around the downtown area.

"No. All I've seen of Manila is Nagtahan."

"Well, then you've only seen the poor part of town," I said. "There's a whole other side of the Philippines you've never seen! Let me show you some of my favorite parts of the city while we wait for your passport. It will be better than sitting in this muggy office building. We can start with lunch. My treat."

We went out for a nice lunch in one of Manila's best hotels. I then took her on a whirlwind tour of the city. It was our first date, only Kristyn didn't know it at the time!

We enjoyed our time together in the city. The hours passed quickly as we talked and took in the sights. I was on cloud nine, so happy to finally be spending time with this beautiful girl that I lost track of the afternoon.

"Kel, what time is it?" she asked as we leisurely walked down a busy street.

"It's after four, why?"

"Four? The passport office closes soon!" She exclaimed.

"I totally lost track of time." I said. When we returned to the embassy it was closed. "I guess we'll just have to come back here tomorrow." I felt bad about the passport, but thrilled it meant we could spend more time together. We returned again the next day and got her extended passport.

Too soon Kristyn's outreach came to an end and she was packing her bags, getting ready to attend the SOE in Hawaii. Every time I thought of her leaving, a bigger lump formed in my throat. We had spent only a few afternoons together. I knew how I felt about her, but she didn't know. "Lord, what do I do?" I wondered.

I stopped by the house she was staying at in Nagtahan the day before she left. "I came to say goodbye," I told her, "and to ask, if it's okay with you, if I write you while you're in Hawaii at the SOE."

"Sure," she smiled weakly.

"I'll miss you," I admitted.

It was only later that she told me she didn't really want me to write, that she just didn't know how to say no. I'm thankful to this day that she didn't know how to say no! When she got to Honolulu I started to send cassette tapes to her, talking about my life and what was happening in the Philippines. I wrote letter after letter.

For quite some time Kristyn didn't write back. I waited for a letter or a tape to arrive in the mail. They never did. "I guess she doesn't feel the same way about me, does she, God?" I prayed, discouraged. I decided to lick my wounds and to no longer pester the girl. As painful as it was, I decided to take the hint and release the relationship back into God's hands.

After a period of silence from Kristyn that felt like years to me, a letter arrived in the mail.

Dear Kel,

How are you? I know it's been a long time since I've written you. To be honest,I didn't know what to do.

I devoured the letter, so pleased to finally hear from her. Little did I know Kristyn was unsure what to do about me. She detailed in the letter that when she arrived in the Philippines, she was in a relationship with someone else. She said it had ended right as her DTS was about to begin.

I went to the leaders of the school in Honolulu and told them about this guy named Kel Steiner in Manila who was writing me all these nice letters. But you see, Kel, I've been holding on to the other relationship even though the other guy had broken it off. I've been asking for council about what to do.

I could hardly read her cursive fast enough. She detailed that, as her leaders listened to her story and prayed, it became clear to them that Kristyn needed to let go of the former relationship completely.

I hooted out loud!

Later she told me that on that very day it was like God took a pair of scissors and cut the ribbon, bringing her to the finish line of the other relationship as it were, and her feelings for the other man completely fell away.

And when the feelings for him fell away, I started to wonder why there were no more letters arriving in the mail from you. I've realized since being in Hawaii just how much I've come to love receiving them. I hope you'll start writing me again.

It was all the encouragement I needed! We wrote. I sent cassettes. We told each other a lot about ourselves. We got to know each other despite being separated by all those miles of ocean. We continued to write and send cassette tapes all throughout Kristyn's SOE: three months in Honolulu and a three-month field outreach to Mexico. When her school was finished, she returned home to Illinois to work.

By this time, it had been over a year since I had seen her. It was time to make an important trip. I visited her at her home, and on that visit, I said out loud the words that have transformed my life ever since, “Kristyn, will you marry me?”

“Yes!” she said, the same smile alighting her face that can calm my whole world.

I was over the moon. I headed back to Manila with a silly grin. My whole life was about to change for the better. We were married three months later in 1985.

Chapter 34: Family Changing Ministry

Kristyn burst into the office. The look on her face said it all. She was shock white, like she'd just witnessed something horrendous.

"What happened?" I asked, standing up from my desk, startled from the quiet afternoon of work at the office. "It's John-Michael," she said. "Lita found him face down in the pool." Kristyn choked on her tears. She was out of breath and could hardly speak.

The blood drained from my body. "No!" I yelled out loud in the office as my world started to spin out of control. I quickly ran around to my wife and wrapped my arms around her. She had limped into the room, still in pain from a C-section she had only two weeks earlier for our newborn baby girl, Karissa.

"Lita was in the kitchen, helping with the meal when she heard the splash. I was watching Karissa in the front of the house," Kristyn went on. She said she thought it was probably just a cat or some other animal that had fallen into the unused swimming pool. The pool in the backyard of the house we rented was full and overflowing with monsoon rains. We regularly had to fish out dead things from the water.

"Lita was chopping up vegetables and I don't know how long it was after she heard the splash," Kristyn continued, "but she decided to go and take a look in the back yard. That's when I heard the scream, Kel, a horrific scream. I ran as fast as I could." She touched her hands to her stomach, wincing in pain.

"John-Michael," I whispered.

"I got there as fast as I could to find out where the scream was coming from. I thought a burglar had broken into the house, and had stabbed her or something. All I could think of was the Shellings. I just thought, Oh no, Lita's been attacked!" She paused for a moment and seemed to be reliving the moment.

"But when I got there I saw why Lita had screamed. John-Michael was floating face down in the pool! All I could see were the soles of his shoes on the surface of the filthy water. His little body drifted toward the center of the pool." My legs nearly buckled and I felt as if I'd been punched in the gut.

"What did you do?" I asked, closing my eyes, not wanting to hear anymore and unable to shut out the horrible image from my mind.

"Lita jumped in, but she's so small and not a good swimmer. The pool was full and overflowing of monsoon water, so it was even deeper than usual and there she was, struggling to keep her head above water. I was witnessing a terrible struggle occurring in the pool. Lita, she would sink down to the bottom then push herself up to the surface of the pool to grab a mouthful of air, and before sinking again, do her best to push John-Michael to the poolside so I could grab him. I could hardly bend over because of the pain, but when he got to the edge, I was able to reach at John-Michael and pull him from the water."

"Oh God this is too terrible!" I exclaimed.

"I got in the taxi almost right after, I had to come to tell you."

"I can't believe we've lost him," I cried aloud, overcome with grief.

"What?" Kristyn asked. "No! Oh Kel, he's not dead. He held his breath that whole time! He's alive. He's okay!"

He held his breath! My little fourteen month-old son. He was alive! I thought I had lost him. I thought his life was taken from us. I thought Kristyn had come to tell me the worst news a parent could ever hear, but she had come to ask me to please come home. She was in so much shock from the terrible experience, I put my arm around her and we both hurried home. I wanted to see my son for myself. And there he was, running around as if nothing had ever happened. He had miraculously held his breath for an extraordinarily long period of time. When he was pulled out of the filthy monsoon-filled pool, he spluttered and gulped and coughed out a little water. Other than that, he had no ill effects whatsoever! God had truly spared John-Michael's life, and the two of us were left feeling as though we had received our young son back from the dead.

"Thank you, Father, for your loving hand of protection over your little ones!" I prayed, aware of my smallness and how much grace I and my family would need to continue to walk forward pursuing God together in ministry. The sobering reality of my new life as a family man on the mission field was seared onto the forefront of my mind.

I had not been responsible for another person like I was now responsible for my family. Undeniably dependent on God, I began to pray: Help me Father to always put you first, then my family, and finally my work as a missionary.

Chapter 35: Called to Europe

In 1989 our family returned to New Zealand for a sabbatical. Josh, our third child was born during our stay in New Zealand. How my young family had grown! What Kristyn and I anticipated would be one year in New Zealand, however, turned into seven.

During that first year back in New Zealand, a friend of mine suddenly died. He had been pastoring a small Assembly of God church in Te Puke, famed as the Kiwi Fruit Capital of the World. The overseer of the church asked Kristyn and I to pastor the congregation until it found a permanent minister. Later, at the end of our time serving the church, Asian Outreach asked us to become the directors of their ministry. Asian Outreach is a faith-based ministry, similar to YWAM.

As with any big decision that affected our family's future, Kristyn and I got on our knees and prayed about God's plan for us as a family. "Show us, Father, what you would have us do." And, as He always is, God was faithful to give us a clear confirmation that it was a ministry we should partner with.

"There's just one more thing we need to do before we go ahead and commit to work with Asian Outreach," I told Kristyn.

"What's that?" She asked.

"I need to honor Loren and Darlene and ask for their blessing to go on leave from our work with YWAM." So I asked Loren, since he was a spiritual overseer in my life, for his blessing. Loren was gracious and gave his approval.

I enjoyed my work with Asian Outreach. It was my job to travel all over New Zealand and beyond, sharing about missions. During one particular conference in Hong Kong, I connected with the director of the work in Norway, Eivind Froen. We became fast friends. He also happened to be a fellow YWAMer. Before joining Asian Outreach, he was the founding national director for YWAM in Norway. During the conference Eivind and his newly appointed director, Toby Killerud, invited me to speak and teach about missions in Norway —to help raise awareness and money for the world wide mission movement.

Norway

"God is at work all over the world," I would share as I traveled to Norway for the first time, late in 1996. I went into churches to speak with congregations about what missions is like, sharing from my experiences in Asia and the South Pacific, and to talk about past moves of God.

I was given an invitation to speak to a small Lutheran church during my whirlwind tour of Norway. It was a state church, and the old priest took considerable pride in the building.

"This church is almost eight hundred years old!" he exclaimed as he showed me around the building with a twinkle in his eye. "It's an architectural treasure with a rich history," he said, beaming with the pride as he took in the building. It was indeed an incredibly well preserved church building, especially since it was constructed completely out of wood. And yes, it was still in use. My fascination with the church, though, had its limits. It did not match the fascination of the priest.

All the history was interesting. The building was beautiful. But I was more interested in the people in the building. They were very few. There were only a handful of people sitting in the two rows, at the very back of the church. There were no children, no teenagers, and no families.

As I made my way up to the pulpit, a negative attitude about there being so few people was brewing within me. What am I doing here, when I could be back home in New Zealand speaking to young people in Faith Bible College who are hungry for God's Word? I asked myself as I made my way down the aisle, past row after row of empty pews. As I was thinking the thought, a vision burst into my mind—it was like the most vibrant colors of paint were thrown against a blank canvas. In my mind's eye I suddenly saw a picture of the church filling up with young people. Each row was filled with teenagers and young adults, eyes wide, expectant, reaching out their hands to Heaven. They were filled with the Holy Spirit and running out of the church to go preach the gospel. It was a powerful image, impossible to forget.

When I returned home to New Zealand after my trip, one night I woke in the small hours of the morning. I bolted upright in bed. Looking over at the alarm clock, the red numbers blinked as they did every hour. It was 4:00 AM.

The hair on the back of my neck stood on end as I heard a man's voice: "The vision is of the Lord. Run with it."

I looked around the room. Kristyn was peacefully asleep beside me. No one else was in the room.

Again I heard the voice. "The vision is of the Lord. Run with it."

Chills went up and down my spine. The voice repeated these words once more. Three times I heard this. Each time was louder than the first, and more emphatic. After the third repetition, that same vision I had in Norway while walking past the empty pews at the Lutheran Church, instantly recurred. The old wooden church was empty at first, and then suddenly, it filled with hundreds of young men and women. They reached their arms to Heaven and accepted Christ in their hearts. The Holy Spirit swooped into the place with power, filling all the young people with the power of God. These new converts then left the church to preach the gospel.

God had spoken clearly to me. When the moment passed and my heart stopped racing, I put my head back on the pillow and went to sleep. In the morning I told Kristyn about the vision.

"What should we do?" she asked.

But we knew what we needed to do. Together we asked God about the vision. He clearly told us to go to Norway. At the time I was Dean of Mission at Faith Bible College in Tauranga. By the end of the week, I had resigned my position. In December of 1996, my responsibilities were completed. As a family, we packed up our things, and headed to Europe.

Chapter 36: On the Move

I wrote to the regional director for YWAM in Northern Europe, Alv Magnus. He was based at Grimerud, which is north of Oslo. I recounted to him my vision of the church in Norway. With reservation, he invited me to come to the Grimerud base. "I'd love to have you," he said. "It's just I don't know exactly what role we can find for you to fill."

"I don't know either," I admitted. "All I know is that the Lord has called us to be in Norway, so we're ready to come in obedience." That was good enough for Alv.

We were based there for a year. All throughout that time I went across the country, speaking to Norwegian churches about missions.

As the year went on, I kept hearing people talk about a place called Restenas. I found myself more and more curious about this place the more I heard, especially because the conversations seemed to relay ever worsening news.

One day I finally turned to a friend on the Grimerud base and asked, "What is this Restenas that everyone keeps talking about?"

"Restenas is the main base in Sweden. It's YWAM's largest facility in Europe," my friend said. "It's located a little north of the city of Gothenburg."

The ministry there was going through a very difficult time, and all the reports we heard were negative. Still curious, I started to ask people what it was that was causing trouble. No one really knew. Very soon I would find out for myself.

I made arrangements to go to White Russia, to a city in Belarus. I was supposed to attend a conference there, and as I was preparing to leave my family, I became aware of a prayer summit that was going to be held in Restenas the same week. Something about the place had my curiosity piqued, so I told Kristyn I didn't think I should go to Belarus. Instead, I believed I should be going to Restenas.

We prayed. It was decided: I would attend the prayer summit in Restenas. I had planned to go to one place, but suddenly had felt an inexplicable need to go somewhere else. At the time, I didn't fully realize it was another U-turn in my long-term plans. It was just like that moment on the road, all those years ago, when I scratched my plans on the way to Angeles City, and headed to the Air Force Base instead. I packed my bags and left for Sweden. Soon all of my plans were out the window. This unexpected change in course was the first step into a whole new adventure.

Restenas

Despite all the negative reports, Restenas was a beautiful property. It sat on one hundred acres of land—half of it farmland, and the other half natural forest. A hundred years earlier the property had belonged to a Christian orphanage. The buildings sat upon scenic land, which sloped down to a beautiful bay.

It seemed there was endless space for activity. There was room for sports and games, and plenty of buildings to house students, staff, and conference guests. It was an ideal facility for a YWAM base. After its orphanage days, it was converted into a prestigious boarding school for the children of Swedish government officials. In those days, they received the best education possible. At a later point, the government built a state school on the property for some four hundred and fifty kids.

But the years had not been kind to the place. Its buildings stood, or perhaps drooped, in a significant state of disrepair. The grass had grown up wildly all over the land. It was so long it bent over in long wisps, overgrown and unruly. There were no fences anywhere along the property. Some buildings had holes in their roofs where water trickled in, soaking the wood so that ceilings and floorboards were rotting.

Engineers suggested bulldozing these once prestigious school buildings. They were considered beyond redemption. The floor of the full size gymnasium, with its basketball court, had buckled and was no longer useable. The whole property had the feeling of a "has-been." The place was worn down, decrepit and sad.

It must once have been a magnificent sight to see, the property that educated Sweden's elite. However, when I first saw Restenas it was very sad looking and terribly rundown. Only a handful of staff occupied the whole property. Was it any wonder they were discouraged? It was impossible for so few staff to oversee the ministry they were attempting, and to care for the place at the same time. In a word, the place was distraught.

Leaders from all over Europe, and from throughout YWAM, had gathered in this once vibrant location to meet and pray. Darlene Cunningham was there, and spoke that morning at the prayer summit. An English prophet named David Saunders had come as well. During the morning meeting he gave a very strong word to the group that had gathered. “It is time to draw a line in the sand," he said. “This is the end of the trail.”

The clear inference was that YWAM needed to make a decision to either keep Restenas, or let the property go before the prayer summit concluded. It was a strong word to the mission; the European Leadership Team had some decisions to make.

When the session concluded, I went outside. The mood was somber in the prayer gathering. It was a serious time. But outside the sky was blue and bright, and the sun was shining. Despite the mood, I felt cheery. I waited on a step outside the building for Darlene. I hadn’t seen her in years and was looking forward to speaking with her.

Darlene walked out of the doors and looked up at the sky. She was her usual warm and pleasant self. We started to walk, catching up as we went, exchanging details about family and our work. When the conversation turned to the prayer summit, Darlene looked over the property and sighed.

“What a place. It really is beautiful, isn’t it?” Like me, she saw what it could be, not what it currently was.

“It sure is,” I mused. “Any leader would give his right arm to have a facility like this to train people for the mission field, don’t you think?”

Darlene smiled in agreement. Then a seriousness overtook her. We were both aware the leadership of YWAM Sweden had decided they no longer wanted the property. It had become too much of a burden. Managing the property had been a struggle that had gone on for years—they were behind on mortgage payments and had burnt out under the stress. If no solution could be found, the property would be turned over to the bank and liquidated.

Darlene watched me for a moment with a knowing look, quietly thinking. "We'll be meeting with the European Leadership Team about it this afternoon. At this juncture they just don't know what to do with it."

At the European Leaders' meeting that afternoon the decision about Restenas' future was to be made. What were they to do? No one wanted to take over the kinds of debts and responsibilities that came with Restenas. Darlene had caught the interest in my words that morning and asked the other leaders, "Why don't you ask Kel and Kristyn Steiner if they would consider taking over the leadership of Restenas?"

It wasn't like we were at the top of some list or anything. There was no list! When the leadership approached me with Darlene's suggestion, my heart's immediate response echoed the Old Testament hero's words: "Give me this mountain!" I tried to hide my surprise and excitement by saying,"Thank you, Kristyn and I and the family will pray about it and get back to you."

As news spread that we were considering a move to Restenas, numerous people offered advice—suggesting I reconsider. "Do you want to fail by taking on that base?" They'd ask. But every time well-intentioned people told me not to take it on, they just stacked the wood higher on my fire!

I left Restenas excited. A flurry of dreams and ideas about how to initiate changes on the base and turn it around were alive in my head. When I reunited with my family in Norway to share with Kristyn and the children about the offer to take on Restenas, all three of the kids burst into tears. They had already been uprooted from their home in New Zealand, left all their friends, and were only just getting adjusted to Norway. They didn't want to move again.

"How about this for a plan?" I suggested. "Let's take a train to Restenas as a family. It will be a fun trip. We'll go to Sweden for a few days and check it out. You can see if you like it for yourselves. Then we'll pray about it as a family." This seemed a reasonable idea. So we took a train trip, and stayed for about a week at Restenas. At the end of our week, the kids were thoroughly impressed that it was the right place to go. We made the move to Sweden as a family.

Chapter 37: Turning Restenas Around

A lesson I learned as a young boy in New Zealand about not giving up in the middle of a struggle served me well in my time as leader of the work in Restenas.

I can still hear my father now, shouting "Stop that pig, stop that pig, Kel!" It was market day and dad was taking a dozen or so pigs to the local farmer's auction. He and my older brother, Ray, were herding the pigs up through a narrow race toward the loading bay. From there they would be trucked to the auction. I wanted to be up front with my dad and Ray, driving the pigs forward to get them loaded; but he told me to stay back in the race in case one of the pigs broke loose. If that happened, it was my responsibility to stop the pig from making its great escape.

Hearing dad's voice jolted me out of a daydream. I looked up. There, barreling toward me, was a runaway pig, running for its life. "Don't let it get past you!" my dad shouted while following in hot pursuit.

The pig was at top speed, careening from side to side down the race toward me. I began to yell, waving my arms in attempts to frighten it and turn it around. But freedom was on its mind. It flew at me with no intentions of being stopped or caught. I braced myself for impact. I was about to full on tackle this pig!

When it was close enough, I launched myself at the determined creature. I grabbed hold of it, wrestling it until I had it by its back legs. Incredibly strong and motivated to get free, the pig struggled wildly to get under the nearby fence. I held on for dear life. It ferreted its snout under the fence and began inching its way under. My eardrums nearly burst under the deafening, inhuman squeals of the animal. Even above the cacophony and tumult, I could still hear my dad yelling, "Don't let it go, don't let it go, Kel!" as he ran to help me.

I held on, but the pig was slowly winning. One of its leg's broke free—desperately, I grabbed its tail. I had one hand clasping a back leg and the other on the tail. I couldn't hold onto the other back leg anymore. It was half way under the fence now, so determined to get to the open field and freedom. I had to let go of the last leg. In a last ditch effort, I clung to that pig's tail with both hands, and pulled for dear life! Just as my dad arrived, I flew backwards and the pig charged forward, squealing out into the open field.

I looked up at my dad, expecting him to scold me for losing the pig. I held out my hand and said, "Look dad, I didn't let go, I didn't let go!" There in my hand was the pig's tail. In its frenzied struggle for freedom, it had given up its tail, root and all. How could my father be mad at me for letting the pig get free when I had the evidence of my struggle!

In some ways, the challenge facing us in Restenas was not unlike my struggle with the pig all those years ago. The base was about to slip out of YWAM's hands—we needed all the energy and faith we could get to keep it from escaping us. Even beyond that, we needed the intervention of our able and loving Father.

We arrived as a family at Restenas in 1998, full of enthusiasm and faith for our time there. The reality on base, however, was pretty grim. It was do or die. And by the looks of things in the natural realm, Restenas would surely die.

We had been in Sweden only a couple of months when the manager at the bank—the overseer of our bank files—came to see me. He wanted Sek 400,000 (approximately $52,000 USD) by Christmas. It was already approaching the end of September! This was a huge reality check. I tried to talk to the bank manager about our plans and vision for the base. He was in no mood to listen. He cut me short, and rubbing his fingers together, indicating all he wanted was money.

"I don't want to hear any plans, visions or strategies," he said curtly. "I just want money." With that, our conversation ended. He left me to ponder our dire circumstances.

I went to my staff to tell them what he'd said. Most of the staff was so discouraged and depressed to hear this. They looked at me, all the color gone from their faces, and asked, "What are we going to do?"

"We'll do what we always do," I said. "Pray." So pray we did.

Before coming to direct the Restenas base, I knew no Swedes. At the very start of my time there I asked the Lord, "What do I do? How do I grow this base?"

He said very clearly to me: Through your relationships, son.

After hearing this, I went through my address book and I began to make phone calls. I wrote letters, and sent cassette tapes with appeals for help to friends around the globe. I encouraged our staff to write on the chalkboard all the names of individuals they thought might be willing to come and serve with us. That list was prayed over regularly and one by one God began to draw those people to come and help us, our staff grew steadily.

Because YWAM Restenas did not have a good reputation, Swedes didn't come at first. As a result, it was a very international group of people that made up the base. The process of turning Restenas' reputation around unfolded because of the great team God gave me to work with.

Mark and Hanelle Erickson, and Mike and Lotta Stevens were part of the original small team I inherited. Mark was a faithful intercessor who committed to pray eight hours each day—a commitment he fulfilled all the years he stayed. Mike was a fellow Kiwi from New Zealand, a former banker who spearheaded the finance department. He could speak Swedish, an enormous asset to the team, as our financial dilemma required that we work with the government in the Swedish language. His Swedish wife, Lotta, did all the translation work required for government documentation. Soo Hong and his wife, Sonny, were pillars of strength; they gave excellent leadership to our growing Korean staff and schools.

Another YWAMer, on the original team I inherited, and the second Swede I came to know, was Leif Isaacson. Leif was a hard, diligent worker, and unbelievably creative. We needed to fix a tiled roof on a five-story brick building that was in disrepair. The cost to bring in professionals to do the job was simply out of the question. We had no money. Leif went into our forest and cut down a number of tall hard wood trees. He stripped off all the branches, and had them looking like giant telegraph polls. He used these polls to build a towering platform, five stories high!

Leif used this platform to stack the tiles he removed from the leaking roof. After replacing the rotten struts where the tiles had been, he returned the original tiles back to the roof. The incredible structure he'd built using the polls was then moved to the next spot. He worked his way across the entire width of the building like that. The job was completed without a hitch, and cost us virtually nothing. Leif saved us thousands of dollars—he also fenced the whole property and built us a hay barn.

Later Philip and Terry McCurley from South Carolina joined our team. Philip became our maintenance manager. He was another multi-talented man who worked tirelessly on projects all over the property. There was a huge amount of upkeep to do, and he was perfectly suited for the job. With all the improvements going on, it wasn't long before the property began to take on a whole new feel. (Philip later became our school training director).

"The atmosphere has changed," visitors commented. They noted that something wonderful was happening at Restenas, it was like a breath of fresh air had swept across the place. It felt alive and new.

Mark and I prayed regularly from the start that God would shake us until only those who truly wanted to be there—whom God wanted to be there—remained. It was very important, if we were to turn things around, that we had only people fully committed to the new vision and leadership. God heard us. He brought new staff to support the work, and directed others to leave to join ministries elsewhere.

We decided, at Mark's suggestion, that we hold a prayer vigil for a month: twenty-four hours, round the clock, for thirty days. We agreed. All of us committed to regular prayer shifts so that no hour of any day, all month long, went without prayers for God's direction and grace, His intervention, and His wisdom. Boy did we pray!

"Lord you know our desperate need," we cried, "in your grace and faithfulness, please provide!"

At the end of the month we gathered as a community to hold a time of praise and thanksgiving. We did this from sun down to sun up to thank God for the miracle he was about to do. We didn't have a single cent toward the 52,000 USD we needed to cover our debt and secure ownership of the property. We put our faith on the line, trusting God to work a miracle. We were encouraged to take up an offering among ourselves. We believed it was to be a sacrificial offering. We committed ourselves to give in a way that would cost us. We needed to put our money where our mouths and prayers were!

"I think we need to give the money for the car, Kris," I said after pulling her aside. She took a deep breath and thought about it. Kristyn and I had recently been given a financial gift from friends in Norway, enough for us to buy a family car, which we had been without for two years.

"I know it's something we really need," I admitted, "but I believe God is asking us to give, even if it hurts. We need to be generous."

"I agree," she said, with a smile. The car was something we needed as a family, but our hearts were stirred to give that money toward Restenas. With joy we gave all the money set aside for our car as a seed offering.

Other YWAMers did whatever God spoke to them to do. Everyone, however difficult it seemed, stepped up to the offering plate. We all gave sacrificially out of our own need. When we brought our offerings to the Lord we all laid hands on it and prayed for God to wonderfully multiply it. "Thank you God that you always do more than we can ask or imagine. Please grow this financial seed of faith." We needed to see it grow—and grow quickly! We had plowed the ground, so to speak, with our prayers. And now with our sacrificial gift, we had sowed good seed.

A month went by. Loren came to visit and encourage us. We met together; I briefed him on our progress. I showed him our numbers, and described our financial situation. "Kel," he said, "I think you need to write to other YWAM bases around the world to see if they want to help."

So I did. The end of October was fast approaching. I sent the letters out. By November, with the deadline imposed on us by the bank looming over our heads, no money had arrived yet. All we had was what we ourselves had given in our base-wide offering.

We started to count down the weeks. If the money was to arrive, it seemed it would be at the eleventh hour. It was a suspenseful time. We waited on God. There were only a few weeks left before the bank would liquidate our assets and send us packing.

It was now early December and I had done all I could do. It was entirely in God's hands. I had an invitation to speak to some YWAM bases in South Korea, but I was very mindful that my staff were on edge about our impending deadline. What would they think about their leader going on a ministry trip in the middle of a crisis? So I spoke to the Lord about it. His response to me: What more can you do if you stay, son?

I knew we had done all we knew to do; nothing could be gained by staying. It was all the encouragement I needed. I packed my bag and was on my way. Soo Hong had organized the trip, and he accompanied me around the country. We visited many YWAM bases and large campus gatherings across South Korea. Soo Hong and I became close friends during that time. After I spoke about missions and evangelism in each location, Soo Hong would share for a few minutes in Korean and conclude the service with an offering.

It was only at the end of the trip that I learned what he had been saying at the end of each meeting. When he handed me a check for $16,000 USD at the airport, my eyes nearly popped out of their sockets! "Where did you get this?" I asked in astonishment.

"It's a gift from Korea," he said with a smile. "You may have noticed that at each place you spoke, I shared with the group in my own language before you left."

"Yes," I said. "I just assumed you were thanking your countrymen and sharing with them our gratitude."

"That's not all."

"What did you say?" I asked.

"I reminded them that at the end of the Korean War, after North and South Korea stopped fighting, there were tens of thousands of orphans in South Korea. It was a very difficult time for our nation. But countries around the world offered their help. One of those countries was Sweden. The Swedish government offered for Swedish families to adopt about 30,000 Korean children orphaned by the war. I only reminded our Korean brothers that when Korea had a great need, Sweden offered us help. And so I asked them, now that Sweden had a need, could Korea not offer help?"

"Soo Hong," I said in a mixture of laughter and tears, "I had no idea!"

I was truly overwhelmed. The Koreans gave generously despite the tough economic times they found themselves in. It was the late 1990's when the economies of most Asian nations had just crashed, including Thailand, Malaysia, Singapore, and Korea. Things were so bad that what had been worth one dollar before the markets crashed was only worth fifty cents afterwards.

Many Koreans were out of work. Famous for their mission sending, many Korean missionaries posted around the world were called home because churches couldn't support them. Asia seemed to be in financial chaos. And yet, in the midst of their own huge financial struggles, Koreans reached into their pockets and gave generously to us. I came back to Restenas with about $16,000 USD towards the $52,000 USD we needed. The Korean YWAMers had sacrificially given almost a third of our total need!

More Good News

When I arrived back in Sweden in mid-December, I received more good news. While I was in Asia, bases around the world had started to send their responses to those letters I had mailed out at the end of October, the letters Loren had encouraged me to write. By the time we neared our Christmas deadline we had just over $50,000 USD! We were fractionally short. It was more than enough to secure our loan.

Mike and I went to the bank with a skip in our step. We were excited to see the manager who didn't want to hear about any of our strategies or plans, only "money." When Mike and I showed him that we had virtually all the money, his face went white with shock. He was dumfounded.

"I was sure you wouldn't come up with the money," he said, the color blushing back into his face once he found his voice. He cleared his throat and admitted after an awkward silence, "I've already sent all of YWAM's business documentation up to Stockholm to arrange for the liquidation of Restenas."

He had to eat humble pie by sending new documentation up to Stockholm explaining his error in judgment. What a happy day that was for us, seeing God redeem Restenas financially! It was an incredibly challenging and fulfilling time. We watched God perform a miracle for us, turning what was perceived to be a hopeless situation into a situation of promise.

From that day on the base never missed a single Mortgage payment. We started to run all kinds of schools. In 2003—the last year our family ministered in Restenas—we were running more schools than any other base in Europe. Eleven total were run that year. In 2003 we committed one million Swedish Crowns to base improvements, and another million to our mortgage. Our hope was to have the mortgage completely paid off by September 2010, the thirty-year anniversary of the Restenas base, and YWAM's fiftieth year as a mission's movement. God blessed our year of jubilee when Loren took up an offering of over one million Swedish Kroner. How we praised God for His faithfulness! YWAM now completely owns it.

The chapel at Restenas, virtually empty when we first arrived, was now jammed full of excited young people. They were full of the Holy Spirit, and training for the mission fields of the world. This was not unlike the vision I had seen a few years earlier in my home in New Zealand. Teams of aspiring youth were being sent into the Caucasus, Eastern Europe, Scandinavia, Africa and India. The Lord of the Harvest had truly blessed Restenas. Its reputation was redeemed. The buildings were refurbished. The land was properly farmed and cultivated. Our staff had grown to seventy-five; and the grounds were peppered with young people from all over the world who had come to learn to do mission work. I will ever be grateful to the many wonderful YWAMers the Lord used to help navigate us through those troubled waters.

What a thrill to watch God at work in our lives! And how important, I realized, that it was to commit things to him in prayer. Like a seed that is planted in the ground, buried and shrouded in dark soil, so too our prayers take root unseen.

“See what God does when we pray and trust in Him?” I was able to say after our faith in the provision of God was tested during those first years. “Just as the farmer trusts his seed to sprout, so too should we expect answers to our prayers to burst forth, sometimes after long periods of germination, to change the landscape of our lives.”

Ground once barren and devoid of hope can be transformed into something new. God does more than we can imagine or think while He brings needed change. Restenas serves as an example to us of this very truth.

Conclusion: Saying Yes

My privilege in serving God these many years resulted from a willingness to say Yes to God.

I am inspired thinking about the future. When I look at the world around me, on my frequent trips, I see people in great need. Everywhere. Families suffer and are in great need: whether they live in Asian slums and are picking through the garbage to make a living, or whether they are an unreached tribe in some remote part of the world, or whether they struggle to make ends meet in a turbulent economy in North America. Wherever I am, there are people who have incredible physical and spiritual needs.

Though the needs of the global population are very great, none are as great as the need for them to hear of God's love through His Son Jesus Christ. There are 2.25 billion people who have never heard the gospel. They are alive right now, waiting to hear this Good News. Some might say these numbers are overwhelming. Yet when I look to the future, it is always with great hope. Sometimes people wonder, how that can be.

I am filled with hope because of the emerging generation. If you're part of it, then I'm filled with hope because you have abilities and gifts, and the sacred ability to say Yes to God.

My own life has been characterized by moments so small and unsuspecting that they are easily overlooked. There were moments when my whole life changed in an instant!

I was driving an old beat-up station wagon to minister to sailors and prostitutes and I miraculously overheard a pilot's radio broadcast. I went with a friend to visit a wealthy lawyer's wife, and we were the answer to her fervent prayers: a revival was birthed. I was packing to head out on a trip to Belarus when I learned about a prayer meeting that would change the direction of my life.

A word someone speaks, a prayer shouted in the desperation of a racing heart, a dream in the night, can all be divinely orchestrated moments that interrupt and change the course of our lives. God has a plan to do something wonderful that we can't quite yet see. He needs us to respond in obedience to Him with an unreserved Yes!

I've thought for a long time about how to end this small book. What story should I tell? What words should I write down? It has not been easy.

So I've decided this. If I had the privilege of sitting with you in the room right now where you sit, reading, I would ask you to put down the book. Then, looking you in the eyes, I'd say: Go. Do it. Risk everything! Take the leap of faith.

You will see the impossible accomplished before your eyes as you say Yes God. People will be changed. Lives will be transformed and, marvelously, one of those changed lives will be yours!

When you look back on your life, I hope—no, I pray—that it will be with the same wonder I have experienced, an incredulous surprise that God has used someone as normal as you to do extraordinary things.

You could say No. We can all exercise free will in that manner. Yet I believe it will be the profound joy of your life, a thrilling and fulfilling adventure until the end of your days on earth, should you enthusiastically say, YES GOD!

Ministry Timeline

1970 - New Zealand team departed Asian Circle Trip

1971 - Returned to The Philippines

1973 - Returned to New Zealand

1974 - DTS in Hilo, Hawaii, USA

1975 - Guam

1979 - Saipan

1983 - The Philippines

1989 - Singapore

1990 - New Zealand

1997 - Norway

1998 - Sweden

2003- Colorado Springs, Colorado, USA

2012 - Myrtle Beach, South Carolina, USA

Appendice

Appendix 1: Prayer for Salvation

Father I have sinned greatly against you, against myself, and others. I am truly sorry. My sin stands like a mountain before me. I want to be saved and set free from my sin.

Thank you for sending your Son, Jesus, to Earth to suffer and die on a cross for me. Thank you for raising Him from the dead. Jesus, please be my Savior. Please forgive me my guilt and sin. I want to be free. I want to start my life all over again. I need to be spiritually born again.

I come to you because I believe only you can save me. You said, "Come to me all you who labor and are heavy laden, and I will give you rest." I've lived a selfish life, so I humbly come to you for forgiveness. Please cleanse me and wash me clean. Only your precious blood can do this.

Jesus, I believe You are the Son of God. I love you. I want to follow you all the days of my life—no turning back. Please take my life. I surrender to you today. Thank You Jesus.

Amen.

If you have prayed this prayer, then please either let Kel or a pastor know about your important decision. Write today!

You can reach Kel at kelsteinerbook@gmail.com

Appendix 2: Call to Missions

The world has a desperate need for more missionaries. Young people must hear the cry of the 2.25 billion people who have never heard the Gospel—they must be willing to take the good news of God's love to them. The Great Commission is a message of love, mercy, and forgiveness that God offers to a dying world. Each generation is responsible for reaching its peers.

It is estimated that we need another 300,000 missionaries if we are to have a real chance of completing the Great Commission in our generation.

YWAM is an interdenominational Christian mission organization. In the last fifty plus years, it has grown to around 20,000 volunteer staff serving in over 1600 operating locations in 186 nations. If you have an interest in knowing what's involved in becoming a missionary with YWAM, visit ywam.org and search for a YWAM school in the country where you would like to be trained.

Kel and Kristyn live in Myrtle Beach where they are pioneering a new YWAM base. You can contact them at info@ywammyrtlebeach.com.

If you are interested in having Kel Steiner speak in your conference, training center or church, then please contact Kel at kelsteinerbook@gmail.com.

Appendix 3: Principles for Effective Intercession by Joy Dawson

1. Praise God for who He is, and for the privilege of engaging in the same wonderful ministry as the Lord Jesus. He ever lives to make intercession for them (His own). Hebrews 7:25

Praise God for the privilege of cooperating with Him in the affairs of men through prayer.

2. Make sure your heart is clean before God, by having given the Holy Spirit time to convict, should there be any unconfessed sin.
If I regard iniquity in my heart, the Lord will not hear me. Psalm 66:18
Search me, O God, and know my heart: Try me, and know my thoughts; And see if there be any wicked way in me, And lead me in the way everlasting. Psalm 139:23-24 (ASV)

Check carefully in relation to resentment to anyone.

Notice the link between forgiveness and prayer in God's Word. In Matthew 6:9-12 when Jesus instructs the disciples how to pray He says, *"Forgive us our debts, as we also have forgiven our debtors"* and immediately following the Lord's Prayer He says, *"For if you forgive men their trespasses, our Heavenly Father also will forgive you; but if you do not forgive men their trespasses, neither will your Father forgive your trespasses"* (verse 14).
Again in Mark 11: 25, *And whenever you stand praying, forgive, if you have anything against anyone; so that your Father also who is in Heaven may forgive you your trespasses.*
Now notice the link between forgiveness and faith when we pray: *Whatever you ask in prayer, believe that you receive it, and you will.* Mark 11:24

Then comes verse 25, warning us to forgive anyone who has wronged us: *Take heed to yourselves; if your brother sins, rebuke him, and if he repents, forgive him; and if he sins against you seven times in the day, and turns to you seven times, and says, I repent' you must forgive him. The apostles said to the Lord, 'INCREASE OUR FAITH!' and the Lord said, If you had faith as a grain of mustard seed, you could say to this sycamore tree, be rooted up and be planted in the sea, and it would obey you.* Luke 17: 3-5
Job had to forgive his friends for their wrong judging of him, before he could pray effectively for them. God restored the fortunes of Job, when he prayed for his friends; and the Lord gave Job twice as much as he had before. Job 42:10
Faith works through love. Galatians 5:6

3. Acknowledge you can't really pray without the direction and energy of the Holy Spirit. The Spirit helps us in our weakness; for we know not how to pray as we ought. Romans 8:26

Ask God to utterly control you by His Spirit, receive by faith that He does, and thank Him. Be filled with the Spirit. Ephesians 5:18

Without faith it is impossible to please Him. Hebrews 11:6

4. Deal aggressively with the enemy. Come against him in the all-powerful Name of the Lord Jesus Christ and with the "sword of the Spirit"—the Word of God. *Submit yourselves therefore to God. Resist the devil and he will flee from you.* James 4:7

5. Die to your own imaginations, desires, and burdens for what you feel you should pray. Lean not unto your own understanding. Proverbs 3:5,6
He who trust in his own mind is a fool. Proverbs 28:26
My thoughts are not your thoughts. Isaiah 55:8

6. Praise God now in faith for the remarkable prayer meeting you're going to have. He's a remarkable God and will do something consistent with His character.

7. Wait before God in silent expectancy, listening for His direction. For God alone my soul waits in silence, for my hope is from Him. Psalm 6:5

But as for me, I will look to the Lord, I will wait for the God of my salvation; my God will hear me. Micah 7:7

But My people did not listen to My voice; Israel would have none of Me. So I gave them over to their stubborn hearts, to follow their own counsels. O, that My people would listen to Me, that Israel would walk in My ways! Psalm 81:11-13

8. In obedience and faith, utter what God brings to your mind, believing. *My sheep hear My voice...and they follow Me.* John 10:27
Keep asking God for direction, expecting Him to give it to you. He will.
.I will instruct you and teach you the way you should go; I will counsel you with my eye upon you. Psalm 32:8

Make sure you don't move on to the next subject until you've given God time to discharge all He wants to say to you regarding this particular burden; especially when praying in a group. Be encouraged from the lives of Moses, Daniel, Paul, and Anna, that God gives revelation to those who make intercession a way of life.

9. If possible have your Bible with you should God want to give you direction or confirmation from it. *Thy Word is a lamp to my feet and a light to my path.* Psalm 119:105

10. When God ceases to bring things to your mind to pray for, finish by praising and thanking Him for what He has done, reminding yourself of Romans 11:36:

For from Him and through Him and to Him are all things. To Him be the glory forever! Amen.

A WARNING

God knows the weakness of the human heart towards pride, and if we speak of what God has revealed and done in intercession, it may lead to committing this sin.

God shares His secrets with those who are able to keep them.

There may come a time when He definitely prompts us to share, but unless this happens we should remain silent. *And they kept silence and told no one in those days anything of what they had seen.* Luke 9:36

But Mary kept all these things and pondered them in her heart.

Luke 2:19

Made in the USA
Charleston, SC
01 November 2015